QUANTUM BIOCHEMISTRY
and
SPECIFIC INTERACTIONS

QUANTUM BIOCHEMISTRY
and
SPECIFIC INTERACTIONS

Zeno Simon, Ph. D.
Professor of Physical Chemistry
University of Timişoara

ABACUS PRESS
Tunbridge Wells, Kent

Revised, up dated translation of
the Romanian edition
"BIOCHIMIE CUANTICĂ și INTERACȚII SPECIFICE"
first published by Editura Științifica, Bucarest, 1973

Translated by Zeno Simon and Vasile Vasilescu
Translation edited by John Hammel, Ph. D.

Abacus House, Tunbridge Wells, Kent, England

ISBN 0 85626 087 8

Printed in Romania

To my wife, most gratefully

Contents

1. Introduction

1.1. Problems of biochemistry and quantum chemistry

In this book we will discuss the problem of the relationship between biochemistry and quantum chemistry. Can, indeed, quantum chemistry methods solve the problems of biochemistry; can biochemistry be reduced to the quantum chemistry of some highly complex systems? Can we, for instance, calculate the oxidation rate of glucose by mitochondria (given the chemical composition and topology of the corresponding reacting system) or can we establish the formula of an efficient antimalignant drug, both of them by means of quantum chemistry?

To answer this question, let us pursue the problems of biochemistry. On opening a biochemistry textbook [1] one notices scores of reactions corresponding to various types of reactions of organic chemistry, lots of association and precipitation equilibria etc. However, the reaction centres are, very often, on biological macromolecules; in most cases reactions occur at interfaces of particles (resembling in size colloidal solutions) or membranes where catalysis by enzymes is very frequent. A more thorough reading reveals situations uncommon to organic chemistry: chemical reactions inside living cell form a huge network within which the end product of a reaction either serves as substrate to one or more distinct reactions, or influences the activity of some enzymes involved in other completely different reactions. Later, highly specific reactions are involved: an enzyme catalyses the transformation of a molecular species from among more similar molecular species in the cell, or a certain antibody neutralises only the particular antigen which elicited it and none of the other over one billion antigenic determinants of the antiboby-producing organism.

According to a principle of dialectical materialism [2] life—a highly complex motion—could not be reduced to motions of lower degree which are the object of physics and chemistry; theories developed to describe systems of a certain complexity could not be extended to treating systems of a higher degree of complexity. If quantum chemistry was to be applied to the study of mitochondrial oxidation one would need, at first, an enormous set of initial data describing the system, which is practically impossible to acquire. Secondly, the corresponding quantum chemical equations, even those required by an *a priori* calculation of the lowest energy-level configuration of a protein molecule in solution, are far too intricate to be solved by even the most powerful modern computers (and very likely, by those of decades to come [3]). Thirdly, as systems become more complex, newer properties emerge [4]. Just as on passing from isolated atoms to diatomic molecules, the phenomenon of atomic vibration in molecule, the infrared and combined diffusion spectra appear. In order that dissociation into ions may occur, liquid water is required in addition to diatomic molecules (e. g. HCl). When passing from micromolecules, which feature organic chemistry, to proteins, one finds colloidal properties and cooperative phenomena which cause, for instance, reversible denaturation.

How, then could quantum chemistry be useful to biochemistry? All quantum chemistry allows at present is to account for the values of rate constants of various elementary reactions and their variation entailed by molecular changes and sometimes, to predict these variations semiquantitatively. This is extremely useful to biochemistry which deals with interactions between molecules in special situations (e.g. fixed mutual orientations) which can hardly, if at all, be studied by experimental organic chemistry. Quantum chemistry also allows one to evaluate the energy of interaction between molecules (not reacting chemically) on the basis of some parameters computed for them; this kind of interaction energy determines the association and precipitation equilibria. Thus quantum chemistry can supply us with numerous data regarding elementary reaction processes and association interactions between biomolecules. Quantum chemistry, especially, facilitates the systematisation of existing experimental data on the processes mentioned above; obtaining some regularities starting with a huge amount of such data is also made easier. From data on elementary processes one can work out a good deal about the rates of various enzymatic reactions and the influence of different factors upon these reactions, about interaction processes (e.g. drug-receptor site) and the degree of specificity of some interactions. The specificity of interaction is basically a problem of distinct chemical affinities, for example, that of the enzyme towards the substrate which it transforms compared to the affinity towards other similar molecular species existing in the cell. Evidently, the treatment of these processes may not be reduced to the quantum chemistry of the systems under discussion; however methods of quantum chemistry together with other physical chemical methods and some empirical regularities (these latter are used mainly in simplifications) make up an approach. Similarly, the theory of processes mentioned above is incorporated in the treatment of more complex systems, such as metabolic chains and various aspects of cellular regulation, of course, together with other methods specific to a higher degree of complexity.

1.2. **Reaction equilibria and kinetics**

The sense and time evolution of processes mentioned above are determined by thermodynamics, that is chemical equilibrium and kinetic aspects, respectively.

We will briefly present some problems of chemical thermodynamics [5, 6]. Changes undergone by various systems (that is our processes) imply a decrease in the sum of chemical potentials $\sum_i \mu_i \nu_i$ of participants. At equilibrium this sum reaches a minimum and macroscopically no further change occurs. For a reaction ($\nu_1, \nu_2 \ldots$ are stoichiometric coefficients)

$$\nu_1 A_1 + \nu_2 A_2 + \ldots = \nu_3 A_3 + \nu_4 A_4 + \ldots \tag{1.1}$$

at equilibrium the ratio of concentrations $[A_i]$ is given by the law of mass action:

$$\frac{[A_3]^{\nu_3} [A_4]^{\nu_4} \ldots}{[A_1]^{\nu_1} [A_2]^{\nu_2} \ldots} = K \equiv e^{-\frac{\Delta G}{RT}} \tag{1.2}$$

To be more exact one should replace concentrations $[A_i]$ with activities a_i of the respective substances; the latter are connected to the former by the equation:

$$a_i = \gamma_i [A_i] \tag{1.3}$$

where γ_i is the activity coefficient which depends on the composition of the reaction medium but is generally close to unity ($\gamma \simeq 1$). The equilibrium constant K is given by the free enthalpy of reaction, ΔG (for reactions occurring at constant temperature T and pressure P):

$$\begin{aligned} \Delta G = \Delta H - T\Delta S &= \Delta U + \Delta(PV) - T\Delta S \\ &= \Delta F + \Delta(PV) \end{aligned} \tag{1.4}$$

where ΔH is the enthalpy of reaction (heat of reaction), ΔS the entropy, ΔU the internal energy, ΔF the free energy of reaction and $\Delta(PV)$ is the variation of product of pressure times volume during the reaction. All these reaction parameters can be related to corresponding reactant parameters in the standard state (activities equal to unity) by the equations:

$$\Delta G = \sum_i \nu_i G_i, \quad \Delta H = \sum_i \nu_i H_i, \quad \Delta S = \sum_i \nu_i S_i \qquad \text{etc.} \tag{1.5}$$

In all these algebric sums the stoichiometric coefficients of products (right-hand side of equation (1.1)) are positive while those of reactants (left-hand side) are taken with a minus sign. In condensed systems the free enthalpy G and free energy F are approximately equal. The product ($P.\ V = 1$ atm $\times$ molar volume) is small in condensed substances.

Chemical reactions always advance in the sense of decreasing G of the system but the composition which corresponds to the law of mass action (equation (1.2)) applied for every possible reaction in the system, is only reached if these reactions are fast enough. This is generally true, from all systems studied by biochemistry, only for protolytic reactions and for interactions without actual chemical reactions (such as associations, dissociations). If not all possible reactions are fast compared to the time interval of interest (e. g. the duration of a cell cycle—at least five hours for mammalian cells), the products corresponding to fast reactions are formed. In these circumstances the composition of the system is determined by reaction kinetics and does not correspond to the lowest possible G state. This is the case with most biochemical reactions: from the multitude of pathways (reactions) which could, in principle, cause a decrease of G of the system, only those for which suitable enzymes are present, take place. The macroscopically observed reactions usually consist of a sequence of at least two or three elementary reactions (reactions at the level of direct interactions between molecules, ions or radicals). The reaction rate depends on the concentration of reactants according to equations of the type [5a]:

$$v = \frac{\mathrm{d}[\mathrm{A}_1]}{\mathrm{d}t} = k_v [\mathrm{A}_1]^{\nu_1} [\mathrm{A}_2]^{\nu_2} \ldots. \tag{1.6}$$

or according to more sophisticated equations. The sum $\nu_1 + \nu_2 + \ldots$ is referred to as the order of reaction and k_v as the rate constant. The elementary reactions are monomolecular—the transformation of a particle or bimolecular—the collision of two particles. The rate constant k_v usually depends strongly on temperature according to the Arrhenius' formula:

$$k_v = A\mathrm{e}^{-\frac{E}{RT}}, \tag{1.7}$$

where E is the activation energy, the excess of energy required by particles to react on collision and A is the pre-exponential coefficient. The equilibrium constant is equal to the ratio of the rate constants $\vec{k}_v$ and $\overleftarrow{k}_v$ for the forward and backward reactions respectively; the activation energies are related to the heat of reaction ΔH:

$$K = \frac{\vec{k}_v}{\overleftarrow{k}_v}; \quad \Delta H = \vec{E} - \overleftarrow{E} \tag{1.8}$$

In Eyring's [7] theory of the activated complex, it is shown that the rate constant (for the reaction $A + B + \ldots \to C^{\neq} \to$ products) may be written, if the activity coefficients are taken into account, as:

$$k_v = \frac{\gamma_A \cdot \gamma_B \cdots}{\gamma_C^{\neq}} \cdot k_v^0; \quad k_v^0 = \frac{kT}{h} e^{\frac{\Delta S^{\neq}}{R}} \cdot e^{-\frac{\Delta H^{\neq}}{RT}} \tag{1.9}$$

where k is Boltzmann's constant.

The activated complex represents the complex of particles in the spatial configuration corresponding to the peak of potential barrier. The quantities corresponding to the activated complex are marked with $\neq$. $\Delta H^{\neq}$ represents the difference between the enthalpy of the activated complex and that of the reactants and is approximately equal to the height of the potential barrier, i.e. to the activation energy. The activation entropy $\Delta S^{\neq}$ determines the pre-exponential coefficient:

$$E \cong \Delta H^{\neq}; \; A \cong \frac{kT}{h} e^{\frac{\Delta S^{\neq}}{R}} \tag{1.10}$$

In biochemistry most of the endergonic reactions — the synthetic reactions—extend over several steps. At least one of the steps consists of a catalytic reaction in which adenosine-triphosphate (ATP) plays the role of a coenzyme and it hydrolyses exergonically into adenosine-diphosphate (ADP) and inorganic phosphate. Thus, in presence of adequate catalyst (enzyme), the endergonic reaction with $\Delta G > 0$ is coupled with an exergonic reaction such that the overall process becomes exergonic, $\Delta G < 0$. Meanwhile the activation energy decreases and there is accordingly, an important increase of the biochemical reaction rate.

In most biochemical reactions the activation energy is supplied by thermal agitation (temperature) [7]. However, many reactions are known whose activation energy comes from the absorption of a light quantum—photochemical reactions, such as photosynthesis. In photochemical reactions [5b], which are usually complex systems of reactions, the first step is always the reaction of a molecule in an excited electronic state which dissociates (breaks down) or has a biradical character. A typical parameter for all photochemical reactions is the quantum yield representing the average number of reacted molecules per absorbed quantum. The radiochemical reactions [5b] are produced by high energy radiation; all throughout their path in matter, ionise and dissociate until their energy is dissipated. In their turn, the radicals, ions and electrons, which are the high-energy primary products can generate secondary reactions with other molecules in the system. The characteristic parameter of radiochemical reactions is the radiochemical yield which represents the number of molecules reacted per 100 eV of energy absorbed from the high-energy radiation.

Quantum chemistry allows the calculation, at least in principle, of internal energies U_i and entropies S_i of various molecules and molecular complexes, of activation energies E and activation entropies $\Delta S^{\neq}$ as well as of characteristics of molecules in excited electronic states.

1.3. **Energetic factors — entropy and partition functions**

In quantum mechanics the energy of a system is calculated from the eigenvalue, ε_0, of the energy operator in the Schrödinger equation for the system involved:

$$\hat{H}\psi_j = \varepsilon_j\psi_j; \quad j = 0, 1, 2, ... \tag{1.11}$$

where subscript j stands for the fundamental ($j = 0$) and excited ($j = 1, 2 ...$) states while ψ_j stands for the corresponding eigenfunctions. In practice, the Schrödinger equation is too complicated to be solved exactly, even for ordinary organic molecules. For this reason, quantum chemistry resorts to various methods of approximation of which the most frequently employed at present are the different versions of the molecular orbital method. The energy of a molecule with no conjugated electrons may be calculated in a good approximation from the sum of bonding energies

$$U_i = \sum_{(\mathrm{AB})} K_i(\mathrm{AB}) \cdot \varepsilon(\mathrm{AB}) \tag{1.12}$$

where $\varepsilon(\mathrm{AB})$ is the energy of the AB bond and $K_i(\mathrm{AB})$ the number of AB type bonds in the molecule i. The most useful values of bond energies are given in Table 1. Calculations of resonance energy due to conjugation, of electric charge distribution within the molecule, of the energy

TABLE 1 **Bonding energies [5, 7]**

Bond	*Energy, kcal mol^{-1}*	*Bond*	*Energy, kcal mol^{-1}*
C—H	98.5	O—H	110.6
C—F	105.4	O—H(H_2O)	109.4
C—Cl	78.5	O—O	33.2
C—Br	65.9	O—F	44.2
C—I	57.4	O—Cl	48.5
C—C	83.1	S—H	81.1
		S=O	112.0
C⋯C (benzene)	116.4	S—Cl	59.7
C=C	145	S—S	50.9
C=C	198	N—H	93.4
C—O	79.6	N—N	38.4
C=O	168.7	N=O	103.9
C=O(CO_2)	191.0	N=O(NO)	149.4
C=O(CO)	255.8	As—H	47.5
C—N	69.7	As—Cl	60.3
C=N	135		
C≡N	207.9	As—Br	51.8
P—C	61.6	As—I	33.1
P—N	66.8	As—As	15.1
P—O	92.0	Cl—Cl	58.0
P—O	128.0	H—H	104.2
P=N	99.0		
P=S	91.0		

of the first electronically excited states against the energy of ground state are made possible by the methods of quantum chemistry. By perturbation methods, one can also calculate the energy of interaction between molecules that do not react chemically. One can evaluate the change of activation energy of a certain type of reaction, on passing from standard reactants to a series of other reactants, the latter not too different from the former.

According to statistical physics [5c], the entropy of a molecule may be calculated by the equation:

$$S_i = k \ln W_i \qquad (1.13)$$

where W_i is the number of microstates which make up the macroscopic state in which the molecule is being observed. For example, in the calculation of configurational entropy of a monodimensional polymer in solution, W represents the number of spatial configurations that the polymer may adopt successively as a result of the deforming Brownian motion. Another example is supplied by reactions in aqueous solutions in which the result is the production of a new molecular species, e. g. A + solvent → B + C or A → B + C; an increase in the entropy of mixing of about + 10 cal mol^{-1} deg^{-1} is found for them (see Appendix 1). In general, the thermodynamic functions can be calculated from partition function Z:

$$U_i = RT^2 \frac{\mathrm{d} \ln Z_i}{\mathrm{d}T}; \quad S_i = RT \frac{\mathrm{d} \ln Z_i}{\mathrm{d}T}; \quad F_i = - RT \ln Z_i \quad (1.14)$$

For pure substances:

$$Z_i = \mathrm{e}^{\frac{U_i(0)}{kT}} \cdot \sum_j g_{ji} \cdot \mathrm{e}^{-\frac{\varepsilon_{ji}}{kT}} \quad (1.15)$$

where ε_{ji} stands for the energy levels of molecule i; the energy of the fundamental state is considered to be zero, $\varepsilon_{0,i} = 0$. g_{ji} represents the degree of degeneracy of levels ε_{ji}, $U_i(0)$ the energy of molecule i at 0°K as related to a certain standard (usually the elements in their natural state at 25 °C and 1 atm). For a solution formed of N_1 molecules of type 1, N_2 molecules of type 2 ..., the partition function is calculated by the equation:

$$Z = \frac{(N_1 + N_2 + , ...,)!}{N_1! N_2! \ldots} Z_1^{N_1} \cdot Z_2^{N_2} \ldots \mathrm{e}^{\frac{U_{\mathrm{int}}}{kT}} \quad (1.16)$$

where U_{int} stands for the energy of interaction of molecules in solution which determines the activity coefficients:

$$\ln \gamma_i = \frac{1}{kT} \frac{\partial U_{\mathrm{int}}}{\partial N_i} \quad (1.17)$$

In this way, quantum chemistry allows the calculation of thermodynamic functions through calculation of energy levels ε_{ji}. It is also worth mentioning that the energy of a molecule can be written in a fairly good approximation as the sum of energies of various degrees of freedom of molecule, while the partition function as the product of partition functions corresponding to those degrees of freedom:

$$U_i = \sum_l (u_{li} + u_{el}^i) \qquad Z_i = Z_{el}^i = \prod_l Z_{li}. \quad (1.18)$$

There are three translational degrees of freedom, three rotational ones (two for linear molecules and none for atoms) and three times the number of atoms in molecule minus six vibrational ones. In addition to these degrees of freedom corresponding to the motion of atomic nuclei, there is the motion of electrons into the molecule to which terms U^i_{el} and Z^i_{el} in equation (1.18) refer.

We are reminded that from a total of N molecules, the number of those on energy level j is given by:

$$\frac{N_j}{N} = \frac{g_j \mathrm{e}^{-\frac{\varepsilon_j}{kT}}}{Z}. \tag{1.19}$$

At a temperature T the average thermal energy (per degree of freedom) is kT and the energy levels are populated up to an energy of the order kT over the energy of the fundamental level of the molecule. At usual temperatures, of the order of 300 K, many of the rotational levels, a few of the first ones for "weak" vibrations and only the fundamental level for "strong" vibration and electron motion, are populated.

1.4. Co-operative phenomena, conformational transitions, allosterism

Co-operative phenomena are specific to systems composed of a very large number of atoms such as the biopolymers. They may be characterised as follows: let there be a process consisting of a large number of elementary steps, i.e. the conglomeration of some molecules from the gas phase to form a liquid spherule or the attachment of four oxygen atoms to a haemoglobin molecule. We talk about a co-operative process if the kinetic and thermodynamic characteristics of an elementary step depend on the number of steps previously completed. For example, attaching a gaseous water molecule to a droplet is more difficult when the "droplet" consists of a single molecule and easier when it consists of 2, 3, 4... molecules (the attraction exerted by 1 or 2, 3... molecules already existing in the "droplet", respectively). Or else, the equilibrium constant for the association of the first oxygen molecule to haemoglobin is lower than those for the association of the second, third and fourth molecules.

One of the effects of co-operative phenomena is that many processes involving biological macromolecules gain the character of an abrupt change with respect to the change of an external parameter (Figure 1.1.). As an example of a non-cooperative process, the dissociation degree α of acetic acid in aqueous solution changes only slightly with tempera-

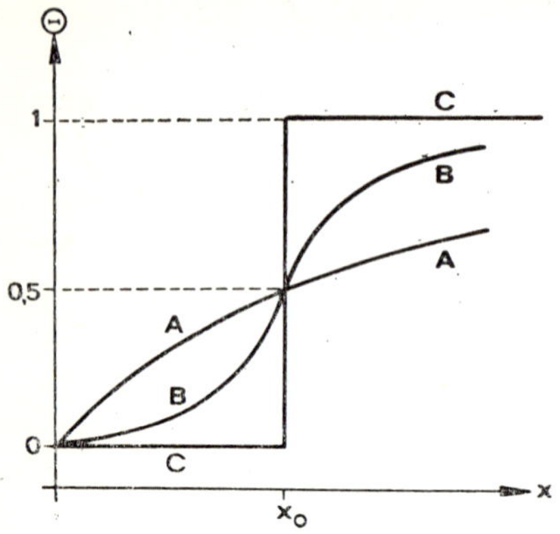

Figure 1.1. The degree of advancement of a process *versus* the change of parameter X
(A — non-cooperative phenomenon; B — highly co-operative phenomenon.

ture between 0 and—50 °C; for a molar solution an increase of ionic strength from 0 to 1 entails increasing α from 0.04 to 0.13 only (curve A, Figure 1.1). The saturation of haemoglobin with oxygen as a function of oxygen partial pressure occurs according to a type B curve; in capillaries $P_{O_2} < X_0$ and haemoglobin discharges to a great extent, while in lung alveoli $P_{O_2} > X_0$ and haemoglobin is loaded with oxygen close to saturation. Curve C corresponds to a phase transition: above a certain temperature (in this case X_0) a phase (e. g. the solid phase) completely transforms into another one (e. g. liquid).

The transitions between various spatial configurations of biomacromolecules have the character of co-operative processes [8]. Thus, at low temperatures, the nucleic acids or the proteins in solution are, at least partially, in an ordered spatial configuration — a state of monodimensional crystal, e.g. the double helix or α-helix respectively. A small temperature increase of up to 10—20 °C about the melting point, entails the rapid decrease to zero of the degree of ordering; the molecule passes into a state of "random coil" — monodimensional liquid. The motivation of co-operative character, of the α-helix → random coil transition of proteins, for example, is the following: the random coil configuration has a higher configurational entropy (equation (1.13)) than that of the "monodimensional crystal" which allows any orientation within the polymer chain-backbone (energetically admitted). The helix appears as a result of the formation of H-bridges between the carbonyl group of an amino acid and the amino group of the fourth neighbour along the chain. On formation of the first H-bridge of this type, of the first helical loop, the spatial conformation of three amino acid residues is fixed. With the formation of the subsequent H-bridges adjacent to the first, on the increase of the α-helix, the spatial configurations of additional amino acids are fixed one by one. Thus the formation of the first loop occurs with a higher configurational entropy loss than the stages involving the growing of the α-helix; the first step advances with more difficulty than the subsequent ones.

A phenomenon encountered in co-operative processes is hysteresis: on returning from the newer value of external parameter x to the former one, the system either does not resume its initial state or it does but on a different trajectory (e.g. on Figure 1.1, the system returns to $x = 0, \theta = 0$, yet not on curve B which it has initially described upon increasing the parameter x). One may offer as an example, the denaturation of proteins: following severe heating, coagulation of the protein may take place and it does not re-dissolve on cooling.

Another phenomenon typical to biological macromolecules is allosterism [9]. The molecule of an enzyme, e. g. has upon it many sites, that is regions with particular affinity towards certain molecular species, which may appear in the "environment". Among these sites, site "A" has a particular affinity for substrate S_A and catalyses its transformation into a product. However, the configuration of site "A" can depend on whether or not sites "B", "C" ... are occupied by other molecules — E_B, E_C ... — called "effectors". Thus in the presence of effector E_B, by its combination with site "B" the configuration of site "A" is modified so as to lower its affinity to substrate S_A. Therefore, effector E_B is a "non-competitive inhibitor" of the catalytic reaction of S_A substrate transformation. Another effector, e. g. E_C, can increase, *via* a similar mechanism, the affinity of site A for S_A or the catalytic capacity of site A, playing the part of an "activator" of the transformation of S_A. Jacob, Monod and Changeux made use of allosterism in explaining the "inductor" or "corepressor" effect usually manifested by substrates and the products of enzymatic reaction chains through repressor inactivation or activation, upon the synthesis of corresponding enzymes. Allosterism is also a co-operative phenomenon since as the allosteric effectors are binding to the enzyme, the binding itself as well as the properties (e.g. catalytic) of macromolecules are altered.

1.5. Structural organisational levels in biomolecules

We deal, in biochemistry, with molecules of different size, ranging from single ions like Na^+, K^+, Cl^- in cytoplasm to complex aggregates of macromolecules which make up the chromosomes and biological membranes [1].

At the bottom of a molecular weight scale we place the inorganic anions and cations from biological fluids, water, monosaccharides, amino acids, the various intermediates of metabolic pathways, etc. The molecular weight of this class does not exceed, generally 250 dalton. Most of the substances used as drugs and vitamins could be included in the same class.

An intermediary position on the same scale, between micromolecules and biopolymers, would be held by lipids, especially the fatty acid esters of glycerol, whose molecular weight (e. g. of tripalmitin) lays within 500—1000 dalton; their marked hydrophobic character explains their tendency to associate in aqueous solutions and produce membraneous formations.

Next higher, on the complexity scale, we can locate biopolymers, proteins, nucleic acids, polysaccharides. In this class, molecular weights exceed 10^4 dalton, the degree of polymerisation is over 100 and biopolymer solutions have a colloidal solution character. The primary structure of biopolymers is very important as, especially so with proteins and nucleic acids, they encode an enormous amount of information. If one defines information as negentropy [10] a polymer with a well-defined primary structure of length N, formed of α types of monomers present in quantities about equal, then:

$$I = S_{\text{conf}} = -k \ln \frac{W_{\text{polymer}}}{W_{\text{randomness}}} = k \ln \alpha^N, \tag{1.20}$$

where W_{polymer} represents the number of microstates of a polymer of well defined primary structure, i.e. a unit, $W_{\text{randomness}}$ the total number of ways one can arrange α types of monomer in a N long chain, that is α^N. Thus a protein made up of $N = 1000$ amino acids ($\alpha = 20$) has, per mole, an informational content corresponding to a negentropy of *ca.* 6000 cal mol^{-1} deg^{-1}. The secondary structure is usually defined as the configurational relations between neighbouring monomeric units along the polymeric chain. One can find "crystalline" secondary structures as the nucleic acid double helix or α-helix and β structure of proteins, and "amorphous", "random coil" secondary structures. A large number of alternative spatial configurations of the polymer chain (or its segments) and accordingly a high configurational entropy may correspond to a certain amorphous structure. The "crystalline" secondary structures are favoured by energetic effects since they are more stable at low temperature. The tertiary structure refers to the global spatial configuration of the whole biopolymer chain. The enzymes and other biologically active proteins—such as antibodies—are often associates of a few (usually 2 or 4) polypeptides. In enzymes, active centres do also frequently contain certain ions or micromolecules involving conjugated systems as "cofactors". To describe the spatial configuration of these "small" associates one resorts to the notion of "quaternary structure". The secondary, tertiary and quaternary structures are determined by the primary structure of biopolymers.

In the organisation of a living cell one can distinguish higher classes with respect to complexity. Thus the next class to biopolymers would be large aggregates of biomolecules like chromosomes, ribosomes, various membraneous formations, etc. The chromosomes consist of DNA chains associated with basic proteins (histones); the DNA double helix is, in its turn, arranged in a "superhelical" configuration. In this complex nonhistonic protein, RNA and ions, especially Mg^{2+}, are also found. The association of biomolecules into these structural complexes seems to imply the formation of covalent bonds through some enzymes.

REFERENCES

1. See for instance E. Soru, *Medical Biochemistry* (in Romanian), Medical Press, Bucharest (1963), Vols **1** and **2.**
2. *Dialectic Materialism* (in Russian), D. I. Danilenko, editor, V-PS Publishing House and OAN-CC-CPSU, Moscow (1961), pp. 139—173.
3. L. W. Linnett, *Nature*, (1971) **234**, 519.
4. J. R. Platt, *J. Theoret. Biol.*, (1961) **1**, 342.
5. E. A. Moelwyn-Hughes, *Physical Chemistry*, Pergamon Press, London (1961), Ch. 6, 20, 21; 5a: *ibid.*, Ch. 22, 14, 5b Ch. 23; 5c Ch. 8.
6. Cl. Nicolau and Z. Simon, *Molecular Biophysics* (in Romanian), Scientific Publishing House, Bucharest (1968), Ch. 1, 2.
7. L. N. Ferguson, *The Modern Structural Theory of Organic Compounds*, Prentice-Hall, New Jersey (1964), p. 48; C. T. Mortimer, *Reaction Heats and Bond Strengths*, Pergamon Press, Oxford (1962), Table 64.
8. T. M. Birshtein and O. B. Ptitsyn, *Conformation of Macromolecules* (in Russian), Publishing House for Science, Moscow (1964) Ch. 9.
9. O. Hörer, *Structure of Biopolymers* (in Romanian), Scientific Publishing House, Bucharest (1970).
10. H. A. Johnson, *Science*, (1970) **168**, 1540.

2. Molecular orbitals

2.1. Molecular orbital theory. Types of chemical bond

Modern quantum chemistry and particularly its applications to biomolecules makes use, to a large extent, of methods stemming from molecular orbital theory [1, 2, 2a]. This theory is based on a theorem of variational calculation which states that if, for a system described by the Hamiltonian operator $\hat{H}$, we possess a set of approximate wave functions, $\varphi_1, \varphi_2, \ldots, \varphi_n$ for the first n levels, we can construct n linear combinations:

$$\Phi_k = c_{k1}\varphi_1 + c_{k2}\varphi_2 + \ldots + c_{kn}\varphi_n: \quad k = 1, 2, \ldots, n \ (1) \qquad (2.1)$$

which approximate more exactly the first n energy levels of the system. These wave functions are obtained by minimising the mean values of an operator F with respect to coefficients c_{kj}. The problem is reduced to solving the system of secular equations:

$$\left.\begin{aligned} &c_1(F_{11} - \varepsilon) + c_2(F_{12} - S_{12}\varepsilon) + \ldots + c_n(F_{1n} - S_{1n}\varepsilon) = 0 \\ &\vdots \qquad\qquad\qquad\qquad\qquad\qquad\qquad\qquad \vdots \\ &c_1(F_{n1} - S_{n1}\varepsilon) + c_2(F_{n2} - S_{n2}\varepsilon) + \ldots + c_n(F_{nn} - \varepsilon) = 0 \end{aligned}\right\} \qquad (2.2)$$

where

$$F_{pq} = \int \varphi_p^* \hat{F} \varphi_q \, dv, \qquad S_{pq} = \int \varphi_p^* \varphi_q \, dV$$

The S_{pq} integrals are also called overlap integrals. The system of secular equations (2.2) can be solved only for n given ε_k values of parameter ε which represent the approximate eigenvalues of the first n energy levels of the system.

Within the frame of molecular orbital theory, the functions φ_q represent atomic orbitals while the more exact functions Φ_k, which describe the motion of electrons about several atoms, are molecular orbitals. The atomic orbitals are approximate wave functions describing the motion of (valence) electrons of a single atom and correspond to a given hybridisation. In this theory, the motion of each electron is considerend to take place in the potential field created by the attraction of positive nuclei together with the repulsion of the other electrons supposed to be in a mean spatial repartition. The matrix elements F_{pq} are considered functions of c_{kj} coefficients (which describe the repartition of electrons) in advanced variants of the molecular orbital theory. The S_{pq} integrals are referred to as "overlap integrals" and are a measure of the interpenetration of φ_p and φ_q atomic orbitals (i.e. the size of spatial regions where both atomic orbitals have large absolute values). If we attach α and β subscripts to Φ_k molecular orbitals, in order to label $+\frac{1}{2}$ and $-\frac{1}{2}$ values of S_z projection of electronic spin we come out with two molecular spin orbitals $\Phi_{k\alpha}$ and $\Phi_{k\beta}$. The electrons of the system (molecule) fill up these molecular orbitals following the order of increasing energy ε_k of the latter. According to the Pauli principle, we can place 0,1 or 2 electrons on each molecular orbital and no electron, or one on each spin orbital. A certain occupation of molecular orbitals by electrons gives an "electronic configuration" which may be described by a determinantal type antiasymmetric with respect to the electron-permutation wave function:

$$\psi \equiv \{\Phi_1^2\Phi_2^2 \dots \Phi_m^2\} = N \begin{vmatrix} \Phi_{1\alpha}(1)\Phi_{1\beta}(1) \dots \Phi_{m\alpha}(1)\,\Phi_{m\beta}(1) \\ \Phi_{1\alpha}(2) \dots\dots\dots\dots\dots \Phi_{m\beta}(2) \\ \vdots \qquad\qquad\qquad \vdots \\ \Phi_{1\alpha}(2m) \dots\dots\dots\dots \Phi_{m\beta}(2m) \end{vmatrix} \tag{2.2a}$$

In this example, if $\Phi_1, \Phi_2, \dots \Phi_m$ denote the first m molecular orbitals of the system which has $2m$ electrons (in the sequence of increasing energy), the electronic ground state of the system is described by the electronic configuration $\Phi_1^2\Phi_2^2 \dots \Phi_m^2$ and the determinantal function given by (2.2a) corresponds to it. The figures in parantheses (1), (2) ... ($2m$) have the meaning of spatial coordinates of the $2m$ electrons while N is a normalisation coefficient.

The probability of finding an electron of Φ_k around the q atom is given by the product $c_{kq}^* c_{kq}$ while the average number of electrons, ρ_q, around q, or more exactly on the atomic orbital φ_q, is given by:

$$\rho_q = \sum_k n_k c_{kq}^* c_{kq} \tag{2.2b}$$

with n_k standing for the number of electrons on Φ_k and c_{kq}^* for the complex conjugate of c_{kq}; for real numbers $c_{kq}^* c_{kq} = c_{kq}^2$. Obviously $0 \leqslant \rho_q \leqslant 2$, that is not more than two electrons can co-exist on an atomic orbital. ρ_q is also termed the charge density on atom q. It is worth mentioning that in the sum (2.2b) the summation index, k, runs through all atomic orbitals. For stable molecules the orbitals are occupied by two electrons each up to a certain orbital energy, $n_k = 2$, while higher orbitals are unoccupied, $n_k = 0$. For radicals, the highest orbital occupied has one electron, $n_k = 1$.

The lower (i.e. more negative) the energy ε_k of a molecular orbital, the higher the energy required to extract an electron of Φ_k orbital, thus the ionisation energy of a molecule parallels the energy of the highest occupied molecular orbital ε_{HO} while the affinity for electrons parallels the energy of the lowest empty molecular orbital ε_{LE}.

Some rules may be advanced which govern the formation of molecular orbitals from atomic orbitals, namely:

(i) Systems of molecular orbitals are made up of atomic orbitals which have non-zero S_{pq} reciprocal overlapping.

(ii) n molecular orbitals are formed from n overlapping atomic orbitals.

(iii) Approximately half of the molecular orbitals (bonding molecular orbitals) have energies lower than the mean energy of the parent atomic orbitals, the other half (antibonding orbitals) possess energies higher than that; molecular orbitals of energy equal to the mean energy may exist which are called non-bonding orbitals.

(iv) Splitting of molecular orbital energies with respect to the energy of atomic orbitals is larger, the larger the overlap integrals (more exactly the interaction energies which parallel the overlaps) S_{pq}, the closer to each other the energies of atomic orbitals and the larger the number of atomic orbitals which overlap.

(v) The distribution of electrons tends to be uniform, the "charge densities" ρ_q, tend to be equal if the energies of the atomic orbitals are equal. In the case of systems of atomic orbitals of distinct electronegativities, the atoms which are more electronegative (more exactly the lower energy atomic orbitals) have higher ρ_q densities. At high differences in electronegativity between atomic orbitals, the resulting molecular orbitals tend to merge (each) into an atomic orbital.

These rules allow interpretation of various kinds of chemical bond in terms of molecular orbital theory. Simple bonds occur between two atomic orbitals, φ_1 and φ_2, from one atom each, which overlap mutually but not with other atomic orbitals. One bonding (Φ_b) and one antibonding (Φ_a) molecular orbital correspond to each bond. The bonding energy

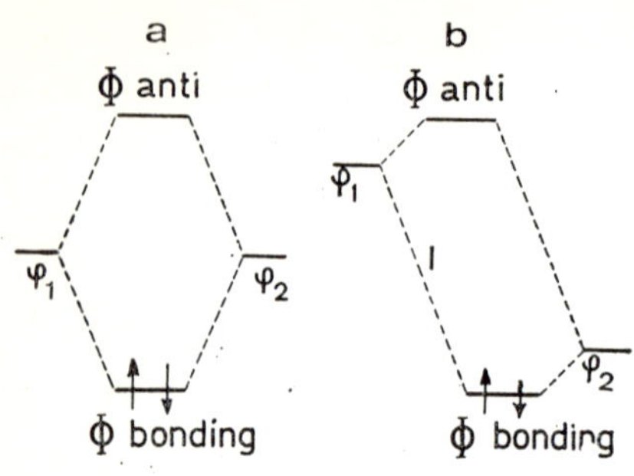

Figure **2.1**. The formation of a simple bond:
(a) covalent bond; (b) ionic bond.

originates in the motion of the two electrons of valence on the bonding molecular orbital, of lower energy than atomic orbitals. If the two atoms have different electronegativities, the bond acquires an electric dipole moment and if the electronegativities are highly different the bonding molecular orbital merges into the orbital of the more electronegative atom so that we get an ionic bond between a negatively- and a positively-charged atom (see Figure 2.1).

If the two atomic orbitals correspond to a symmetric repartition of electron density along the chemical bond, there is a σ-bond. These bonds frequently occur between hybridised atomic orbitals. In their case, neither the overlap integral, S_{12}, nor the bond strength change on rotation of an atom around the bond; the rotation of groups around a σ-bond is, energetically, facile. If there are atomic orbitals perpendicular on bond direction, a π-bond superposes over the σ-bond. The S_{12} integral between the two π-atomic orbitals is maximal if they are parallel; the rotation of atomic groups around the chemical bond leads to decreasing of S_{12}, to breaking of the π-bond and thus requires consumption of energy. There are also atomic orbitals (usually hybridised) which do not overlap with other atomic orbitals — the so called non-participating atomic orbitals — which donate relatively easy their "lone" electron pair.

In conjugate systems, there are chains of orbitals (perpendicular on the plan of molecule) which overlap; excepting the atomic orbitals at the end of the chain, each of the others overlaps with at least two neighbours. The system of molecular orbitals extends all over the corresponding atoms. For example, in benzene, the π-electrons system extends over all the six carbon atoms so that there are six molecular orbitals, three of them bonding which bear the six π-electrons. Another example is suggested by adenine (Figure 2.2); it has the conjugate system formed from the nine atomic orbitals of the ring to which one adds one atomic orbital of aminic nitrogen. The conjugated system contains the electrons of the four π-bonds and the "lone" electron pairs on the aminic and iminic nitrogen

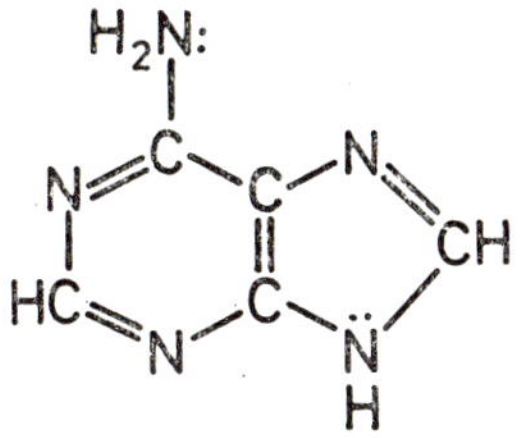

Figure **2.2.** Adenine

atoms. The 12 π-electrons are set on the six lowest molecular orbitals while four molecular orbitals are empty.

In transition metal complexes, since the atomic orbitals of ligands are, in general, appreciably more electronegative than that of the central cation, the molecular orbitals correspond to a rather low degree of electron donation from ligand orbitals to the previously empty orbitals of the central cation. Wherever metal-ligand overlaps occur, the energy of "molecular" orbitals corresponding to ligands is lowered while the energy of orbitals corresponding to the central cation is increased.

The energy bands originate, within the theory of molecular orbitals, in interactions between the orbitals of a large number N of particles (atoms, ions, molecules) placed into the sites of a crystal lattice. The g (molecular) orbitals of energy ε_i of every interacting particle form Ng orbitals extended over the whole crystal, of energies very close to each other. This set of energy levels forms an energy band whose limits are $\varepsilon_i - \delta\varepsilon$, $\varepsilon_i + \delta\varepsilon$; the energies of electrons may take practically continuous values within this interval. The width of this band, $2\,\delta\varepsilon$ is larger the stronger the interaction between particles. Electric conductivity occurs if there is a "cloud" of delocalised electrons, i.e. a band which is only partially occupied by electrons, eventually electrons coming from impurity levels positioned between the highest occupied and first empty band (semiconductors) or electrons promoted from the highest occupied to the first empty band (photoconduction). If the interaction between particles is relatively weak, one cannot conceive conduction bands but a migration of excitation energy from particle to particle takes place (e.g. between aromatic amino acids in proteins, usually placed at fairly large distances from one another).

2.2. **Hückel molecular orbital method (HMO)** [1a, 2]

The molecular orbital method according to Hückel (HMO) has been and is extensively employed in the study of conjugated systems of organic molecules. This method implies a series of very drastic simplifications within the frame of the general molecular orbital method. Experience with HMO calculation, over a long time, indicate that this method reproduces correctly the variations of electron density distribution and the energy levels in large series of molecules not too different from each other. The success of this extremely simple method may be, at least partially, ascribed to its employing parameters standardised against experimental data which, from the beginning, compensates errors introduced by simplifications. For the (sometimes) very large molecules which focus the interest of biochemistry, the HMO method is often the only applicable method. Thus a phthalocyaninic cycle has 40 atomic orbitals participating in a π-electronic system and substitution derivatives of this cycle may easily reach 50 or even more atoms in the conjugated system.

The HMO method may be characterised, against the general scheme of molecular orbital theory, by the following simplifying assumptions:

(i) It deals only with π-electron system. σ-bonds are considered to affect, at most, α and β parameters.

(ii) Each π-electron is considered separately; F_{pq} elements are considered independent of resulting charge distribution.

(iii) The matrix elements F_{pq} and S_{pq} corresponding to system (2.2) of equations are taken as follows:

$$F_{qq} = \alpha_q = \alpha + \delta_q\beta \text{ — proportional to ionisation potential of one electron in valence state of atom } q \tag{2.3a}$$

$$F_{pq} = \begin{cases} 0 \text{ — for non-neighbouring atoms} \\ \beta_{pq} = \eta_{pq}\beta \text{ — proportional to strength of bond between } p \text{ and } q \text{ atoms} \end{cases} \tag{2.3b}$$

$$S_{pq} \begin{cases} = 0 \text{ — for } p \neq q \text{ — so called zero differential overlap (z.d.o.) approximation} \\ = 1 \text{ — for } p = q \end{cases} \tag{2.3c}$$

Herein α_q is the so called Coulomb energy for q atomic orbital while β_{pq} is the exchange energy corresponding to the bond between p and q atomic orbitals. The standard values α and β refer to the $2p_z$ atomic orbital for carbon atom in benzene and to the π-bond in benzene, respectively, which in turn correspond to F_{cc} and $F_{c'c}$ for a $2p_z$ orbital of the carbon the atom in sp^2 hybridised state.

The integrals are negative but the value they are ascribed depends on the magnitude we want to calculate. For resonance energy calculations $\beta = -16{,}5 \ldots -20$ kcal mol^{-1} is recommended while for electronic transition calculations one uses $\beta = -2 \ldots -2.5$ eV ($-45 \ldots -60$ kcal mol^{-1}) [1a, 2].

As an example here are the secular equations and the molecular orbitals for ethylene:

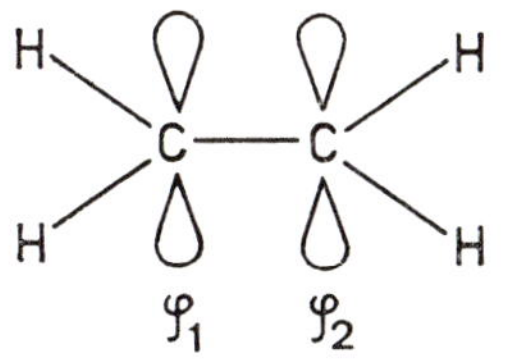

$$\begin{cases} c_1(\alpha - \varepsilon) + c_2\beta = 0 \\ c_1\beta + (\alpha - \varepsilon)\, c_1 = 0 \end{cases}$$

$$\begin{aligned} (\varepsilon_{\mathrm{LE}})\quad \varepsilon_2 &= \varepsilon_{\mathrm{anti}} = \alpha - \beta; \quad \Phi_2 = 0.707(\varphi_1 - \varphi_2) \\ (\varepsilon_{\mathrm{HO}})\quad \varepsilon_1 &= \varepsilon_{\mathrm{bond}} = \alpha + \beta; \quad \Phi_1 = 0.707(\varphi_1 + \varphi_2) \end{aligned} \qquad (2.4a)$$

For benzene

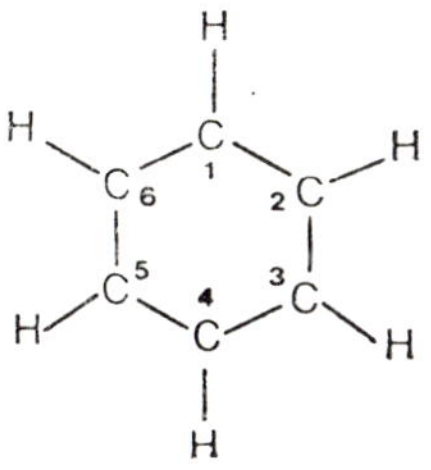

$$\begin{aligned} c_1\beta(\alpha - \varepsilon) + c_2\beta + c_6\beta &= 0 \\ c_1\beta + c_2(\alpha - \varepsilon) + c_3\beta &= 0 \\ c_2\beta + c_3(\alpha - \varepsilon) + c_4\beta &= 0 \\ c_3\beta + c_4(\alpha - \varepsilon) + c_5\beta &= 0 \\ c_4\beta + c_5(\alpha - \varepsilon) + c_6\beta &= 0 \\ c_1\beta + c_5\beta + c_6(\alpha - \varepsilon) &= 0 \end{aligned} \qquad (2.4b)$$

$$\varepsilon_3 = \alpha - 2\beta \quad \Phi_3 = \frac{1}{\sqrt{6}}(\varphi_1 - \varphi_2 + \varphi_3 - \varphi_4 + \varphi_5 - \varphi_6)$$

$$\varepsilon_2 = \alpha - \beta \quad \Phi_2 = \frac{1}{\sqrt{12}}(2\varphi_1 - \varphi_2 - \varphi_3 + 2\varphi_4 - \varphi_5 - \varphi_6)$$

$$\varepsilon_{\bar{2}} = \alpha - \beta \quad \Phi_{\bar{2}} = \frac{1}{2}(\varphi_2 - \varphi_3 + \varphi_5 - \varphi_6)$$

$$\varepsilon_{\bar{1}} = \alpha + \beta \quad \Phi_{\bar{1}} = \frac{1}{2}(\varphi_2 + \varphi_3 - \varphi_5 - \varphi_6)$$

$$\varepsilon_1 = \alpha + \beta \quad \Phi_1 = \frac{1}{\sqrt{12}}(2\varphi_1 + \varphi_2 - \varphi_3 - 2\varphi_4 - \varphi_5 + \varphi_6)$$

$$\varepsilon_0 = \alpha + 2\beta \quad \Phi_0 = \frac{1}{\sqrt{6}}(\varphi_1 + \varphi_2 + \varphi_3 + \varphi_4 + \varphi_5 + \varphi_6)$$

The results of these HMO calculations are listed as energy parameters and as parameters regarding π-electron density distribution [1b]. These energy parameters are:

(i) The energy of molecular orbitals, $\varepsilon_i = \alpha + \lambda_i\beta$, especially the energy of the highest occupied orbital, ε_{HO}, which measures the reducing properties of the π-electronic system, the energy of the lowest empty orbital, ε_{LE}, which describes oxidising properties of the system and the difference $\varepsilon_{LE} - \varepsilon_{HO}$ connected to the wave length λ_1, of the first transition in the electronic absorption spectrum (maximum of absorption band at largest wave length).

(ii) The delocalisation energy, E_D, that is the difference between total energy

$$E_{tot} = \sum_i n_i \varepsilon_i, \tag{2.5}$$

and the energy of π-electrons localised at the parent atoms as well as the resonance energy E_R, i.e. the difference between E_{tot} and the energy of π-bonds, each localised at a pair of neighbouring atoms. Both E_D and E_R as well as their per electron values, E_D/n_π and E_R/n_π are connected to the general stability of the system. E_R is the so called resonance energy; n_π is the number of electrons in the conjugated systems.

(iii) The localisation energy, E_L, the difference between the total energy of conjugated system and the energy of what is left from the system after removing an atom together with 0 or 2 π-electrons (following formation of a new σ-bond).

The electron distribution parameters are:

(i) The π electric charge of atom q:

$$\rho_q = \sum_i n_i c_{iq}^2 \tag{2.5a}$$

(ii) The bond order between atoms p and q:

$$p_{pq} = \sum_i n_i c_{iq} c_{ip} \tag{2.5b}$$

(iii) The free valence index:

$$F_q = C - \sum_{p-\text{neighbours}} p_{qp} \tag{2.5c}$$

The last summation is done over the atoms directly bonded to the q atom and for sp^2 hybridised carbon atom C is 1.732.

(iv) The polarisability of the bond between p and q atoms

$$\pi_{pq} = \sum_{i \neq j}\sum n_i n_j \frac{c_{ip}c_{iq}c_{jp}c_{jq}}{\varepsilon_i - \varepsilon_j}. \tag{2.5d}$$

Subscripts i and j refer to molecular orbitals, p and q to atomic orbitals, n_i and n_j to the number of electrons on Φ_i, Φ_j molecular orbitals; $n_i = 0$, 1 or 2.

As an example, for the ground state of ethylene, the electronic configuration is Φ_1^2, $n_1 = 2$, $n_2 = 0$, $E_D = 2\beta$, $E_R = 0$ (only one π-bond in this system) $p_1 = p_2 = 1$, $p_{12} = 1.000$, $F_1 = F_2 = 0.732$. For benzene in the ground state the electronic configuration is:

$$\Phi_0^2\Phi_1^2\Phi_{\bar{1}}^2, \quad n_0 = n_1 = n_{\bar{1}} = 2; \quad n_2 = n_{\bar{2}} = n_3 = 0;$$
$$E_D = 8\beta, \quad E_R = 2\beta, \quad \rho_1 = \rho_2 = ... = \rho_6 = 1, \quad p_{12} = p_{23}$$
$$= ... = p_{61} = 0.666, \quad F_1 = ... = F_6 = 0.400.$$

An outstandingly important problem for the success of the HMO method is the selection of α_q and β_{pq} as well as of the number ν_q of electrons and the number of orbitals contributed by each atom to the conjugated system. Integrals α_q are usually taken proportional to the electronegativities of the valence state of the atoms. However, they also depend on the number of electrons shared by the atom with the conjugated system and on the (formal) electric charge of the atom. For example, the oxygen atom shares one π-electron with the conjugated system in the carbonyl group $\rangle$C=O, in which case one takes $\alpha_0 = \alpha + \beta$, and 2π electrons in the —ÖR or ÖH groups in which case $\alpha_0 = \alpha + 2\beta$. In the phenolate anion, the formal charge of —Ö$^{\ominus}$ group is —1, $\nu_q = 2$ and one considers $\alpha = \alpha_0 + \beta$, oxygen being less electronegative than an undissociated OH group. In oxonium derivatives O$^{\oplus}$ contributes with $\nu_q = 1$ but due to the formal positive charge the electronegativity increases and $\alpha = \alpha + 2\beta$. Streitwieser [2] recommended a $\Delta\alpha_q$ variations of α_q integral of 1 to 1.5 units per electronic charge unit. The groups bonded to atom q may also influence α_q; Streitwieser also recommended a variation $\Delta\alpha = (0.1 ... 0.05) \times (\alpha_x — \alpha)$ for C atoms bonded to the X heteroatom and $\Delta\alpha = —0.2\beta$ for substitution of H by a CH_3 group, since methyl group is electron donating as compared to hydrogen.

Integrals β_{pq} are taken proportional to the strength of the π-bond between p and q atoms, that is to overlap integral S_{pq}. They depend not only on the nature of *p* and *q* atoms but also on the (formal) type of bond between them (double or single with conjugation in the classical chemical formula), on the coplanarity of groups bonded to *p* and *q* atoms. For an angle θ between the two plans of atoms bonded to *p* and *q* one employs the relation [1d, 2]:

$$\beta_{pq}(\theta) = \beta_{pq} \cdot \cos\theta. \tag{2.6}$$

As an example β_{CC} is considered 1.0β for CC bonds in aromatic hydrocarbons ("half double bond"), 0.8—0.9β for formally simple CC bonds (like: CH_2:CH—CH:CH_2 in butadiene) and 1.1—1.2β for double ethylenic bond (in $H_2C{=}CH_2$).

All handbooks on the HMO method contain tables of α_q and β_{pq} values, generally different from each other. A tabulated selection of such values is given in Purcell and Singer's work [3] issued in 1967; ever since more and more advanced methods have been used preferentially due to increasing computing abilities so that this work practically lists all recommended values of α_q and β_{pq}. Several α_q — β_{pq} sets of values yield good results; this may be accounted for by perturbational considerations and by heteroatoms being usually bonded to carbon atoms in organic molecules. The δ_X and η^2_{CX} values which lead to the same η^2_{CX}/δ_X quotients may be shown to give approximately the same molecular orbital energies and charge distributions. Table 2.1 give the most usual values of α_q and β_{pq} employed in HMO calculations.

The HMO method was initially issued for dealing with conjugated systems in organic molecules. Separation of σ-electrons from π-electron system is justified somewhat for planar molecules whose overlap integrals of π and σ atomic orbitals are zero, $S_{\pi\sigma} = 0$. In non-planar molecules this is not strictly so but the HMO method also applies with good results. The influence of heteroatoms not involved in conjugation may be taken into account through a Δα variation of α_q integral of the conjugated atom bonded to the heteroatom. In the case of hyperconjugation, the influence of CH_3 and RCH_2 groups upon the conjugated system can be treated by several methods [1d, 2]; the simplest considers the methyl group as a heteroatom with $\nu_{CH_3} = 2$ electrons, $\alpha_{CH_3} = \alpha + 2\beta$, $\beta_{C-CH_3} = 0.6\beta$. Most HMO calculations were performed for conjugated systems of organic molecules as spectra and the majority of their reactions are determined by the systems of conjugated bonds (and pairs of non-par-

TABLE 2.1 **Parameters for HMO calculations** [2]

(a) Conjugate systems (number of points, indicates π-electrons per atom)

$\delta_{-\dot{C}-} = 0$; $\delta_{-\dot{N}-} = 0.5$ (pyridine); $\delta_{-\ddot{N}-} = 1.5$ (pyrole);

$\delta_{-\dot{N}-}^{\oplus} = 2.0$ (pyridinium); $\delta_{-\ddot{O}-} = 2.0$ (ester); $\delta_{-\dot{O}|} = 1.0$ (carbonyl);

$\delta_{-\dot{O}-}^{\oplus} = 2.5$ (pyrillium);

$\delta_{-\ddot{O}|}^{\ominus} = 0.9$ (phenolate)

$\delta_{-\ddot{F}|} = 2.5$; $\delta_{-\ddot{Cl}|} = 1.8$; $\delta_{-\ddot{Br}|} = 1.5$; $\delta_{-\ddot{I}|} = 1.2$; $\delta_{-\ddot{C}H_2R} = 2.0$;

$\delta_{-\dot{S}|} = 0.00$ (sulphides); $\delta_{-\ddot{S}|} = 1.0$

$\delta_{-\ddot{P}-} = +0.6$ (pyrole analogue); $\delta_{-\dot{P}-} = -0.4$ (pyridine analogue);

$\delta_{>\dot{P}<} = -0.6$ (tetracoordinated analogue)

$\delta_{C(\ddot{X})} = \frac{1}{20}\delta_{\ddot{X}}$; $\delta_{C(\dot{Y})} = \frac{1}{10}\delta_{\dot{Y}}$; $\delta_{C(CH_3)} = -0.2$

$\eta_{C\overset{...}{=}C} = \eta_{C\overset{...}{=}O} = \eta_{C\overset{...}{=}N} = 1.00$ (aromatic cycles)

$\eta_{C=C} = 1.1$ (ethylenic bond); $\eta_{C-C} = 0.9$ (simple bond)

$\eta_{C-O} 0.8$ (ether); $\eta_{C=O} = 1$ (carbonyl); $\eta_{C-N} = 0.8$ (amine);

$\eta_{C=N} = 1.0$ (double bond)

$\eta_{N=N} = 1$ (azo derivatives); $\eta_{N\overset{...}{=}N} = 1.0$ (diazo group)

$\eta_{C-F} = 0.5$; $\eta_{C-Cl} = 0.4$; $\eta_{C-Br} = 0.3$; $\eta_{C-I} = 0.3$; $\eta_{C-CH_2R} = 0.6$

$\eta_{C=S} = 0.6$ (sulphides); $\eta_{C-S} = 0.6$ (sulphone)

$\eta_{C-P} = 0.9$ (trivalent P); $\eta_{C-P} = 0.9$ (pentavalent P).

(b) σ — systems [6] $\beta_{CC} = \beta_\sigma \cong 4\beta_\pi$

$\delta_C = 0.6$; $\delta_H = -0.2$; $\delta_F = 0.9$; $\delta_{Cl} = 0.5$; $\delta_{Br} = 0.45$; $\delta_I = 0.4$

$\eta_{CC} = 1$; $\eta_{-C-} = 0.34$ (between sp^2AO's on the same atom); $\eta_{CH} = 1.1$

$\eta_{CF} = 0.5$

$\eta_{CCl} = 0.65$; $\eta_{CBr} = 0.58$; $\eta_{CI} = 0.53$.

(c) Transition metals [19] δ_M calculated from χ_M electronegativities by the equation $\delta_M = \chi_M - 2.5$

$\delta_{Sc} = -1.3$; $\delta_{Ti} = -1.2$; $\delta_V = -1.0$; $\delta_{Cr} = \delta_{Mn} = \delta_{Fe} = \delta_{Zn} = -0.9$;

$\delta_{Co} = \delta_{Ni} = \delta_{Cu} = -0.8$.

ticipating electrons). The best treated, from the standpoint of the α_q and β_{pq} parameters are, accordingly, the elements of the second period and the halogens. For elements of following periods and for metals considerably less data are available, especially in cases where the same atom contributes with more than one orbital to the conjugated system. Data for P and S atoms are available in reference [3]. For pentavalent tetracoordinated P best results are obtained with Dewar's model [4], which considers two orthogonal 3*d* orbitals at atom P with interruption of conjugation and $\alpha_P = \alpha_{P'} = \alpha - 0.6\beta$; $\eta_{CP} = 0.9\beta$, $\eta_{PP'} = 0$. The 3*s* and 3*p* orbitals of P are caught, in this case in the system of σ-bonds [5]. HMO calculations were also performed for σ-bonds, with large β_{pq} values for pairs of hybridised atomic orbitals involved in σ-bonds and small β_{pq} values for hybridised atomic orbitals of the same atom (approximate separability for simple bonds) [6]. Some α_q and β_{pq} data for σ-bonds and transition metals are also listed in Table 2.1.

The effect of a substituent may be described rapidly by a perturbational technique. The substitution of atom C subscript *q* — with an atom X produces a $\Delta\varepsilon_i'$ charge of *i* molecular orbital energy while substitution of H of C_q by radical R a variation $\Delta\varepsilon_i''$ given by:

$$\Delta\varepsilon_i' \cong c_{qi}^2 \cdot \delta_X\beta \tag{2.7a}$$

$$\Delta\varepsilon_i'' = \frac{c_{qi}^2 \cdot \eta_{CX}^2\beta}{\lambda_i - \delta_X + \eta_{CX}\mathrm{Sign}(\lambda_i - \delta_X)} \tag{2.7b}$$

Figure 2.3. illustrated π-electron charges, ε_{HO} and ε_{LE} values for four natural purine and pyrmidine bases and for the aromatic radicals of natural amino acids.

2.3. **Reactivity, physical properties and molecular orbital parameters**

Some physical and chemical properties can be correlated with parameters of molecular orbital methods. These correlations are described below for the HMO method but the results may be generally extended to other molecular orbital methods. In the discussion of advanced methods wherever differences from the HMO method appear, special mention is made to them.

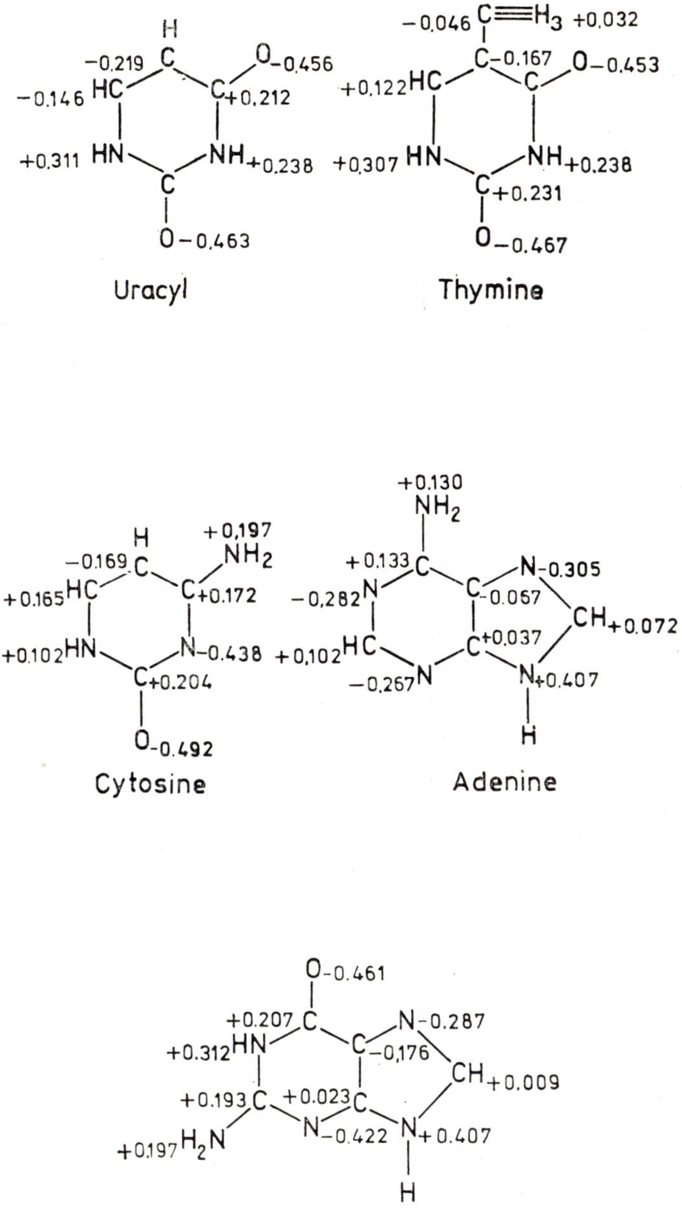

Uracyl

Thymine

Cytosine

Adenine

Guanine

Figure 2.3. π-electron charges, ε_{HO} and ε_{LE} values given by HMO method (according to Pullman [1] Chap. 5).

2.3.1. *Electronic spectra* [1, 2]

The absorption bands which appear in visible and ultraviolet light are assigned to the transition of one electron from occupied molecular orbitals to higher empty ones. Let ε_i and ε_j be the energies of higher and lower molecular orbitals respectively, in which case the wave length of transition is:

$$\lambda_{ij} = \frac{hc}{\varepsilon_i - \varepsilon_j}. \qquad (2.8)$$

The HMO method does not differentiate between transitions with and without change of electronic spin (multiplicity) but usually excited triplet states have lower energies than excited singlet states (which correspond to transition of one electron between molecular orbitals of identical spin). In molecules with large conjugated electron system, the first band at long wave length, usually corresponds to the transition between highest occupied and lowest empty molecular orbitals (the so called $\pi\pi^*$ transition); in transition metal complexes the band corresponds to transitions between bonding and antibonding d orbitals (dd* transitions). Sometimes the first band corresponds to a transition from a non-bonding atomic orbital on the lowest empty molecular orbital ($n\pi^*$ transition). At a shorter wave length one comes accross transitions between orbitals corresponding to localised bonds ($\sigma\sigma^*$ transitions) while in the far ultraviolet region we find Rydberg transitions that imply highly excited valence states of one atom in the molecule.

The intensity of absorption is determined by transition momentum along one of the three axes (x, y, z) of the molecule:

$$(x)_{ij} = \int \Phi_i \cdot \hat{x} \cdot \Phi_j \, \mathrm{d}V. \qquad (2.9)$$

In order that the transition is allowed the $(x)_{ij}$ integrals have to be non-zero, i.e. the product $\Phi_i \hat{x} \Phi_j$ of molecular orbitals involved must satisfy conditions concerning the spatial symmetry of the molecule. The allowed transitions usually lead to high molar absorption coefficients, $\varepsilon_{mx} \cong 10^4 - 10^5$ units (mol^{-1} cm^{-1}). The transitions which are forbidden by the molecular symmetry point group still give absorption bands of lower intensity corresponding to $\varepsilon_{mx} \cong 10 - 10^3$ assigned to the coupling of electron motion with nuclear vibration. For organic molecules, transitions between states of different spin multiplicities, transitions from singlet ground state (quantum spin number $S = 0$) to excited triplet states, $S = 1$, are strictly forbidden.

Once the molecule has reached an excited electronic state, it deactivates by one of the following processes [1]:

(i) chemical reaction, direct dissociation of one bond, or reaction with another molecule; many excited states have biradical character.

(ii) non-radiative transitions — passing to electronic ground state, often *via* an intermediate triplet excited state with transformation of electronic excitation energy into vibrational energy. The resulting excess of vibrational energy dissipates by intermolecular collisions.

(iii) luminiscence — transition to electronic ground state (usually singlet) by emission of one quantum; emission may occur either from the first excited singlet state — fluorescence — or from the first excited triplet state — phosphorescence. The duration of light emission is shorter in the former case, 10^{-8} — 10^{-5} s, and longer in the latter, 10^{-3} — 1 s.

The way the deactivation of the excited molecule takes place depends on the relative rates of the competing processes.

2.3.2. *The bond order*, p_{rs}

This establishes the connection with the length of the respective bond; the shorter the bond (variation interval about 0.2 Å), the higher the dissociation energy. In the vibrational spectra, the band corresponding to the bond shifts towards larger wave numbers with the increase of bond order (variation interval ca. 100—200 cm^{-1}).

2.3.3. *Static reactivity parameters* [1a, 2]

There are parameters referring to the molecule of interest, in the electronic ground state for thermal reactions or in an excited electronic state (usually the first one) for photochemical reactions. These parameters (atomic charges, free valence indexes, etc.) give a satisfactory correlation with reactivity if the activated complex is similar to the initial molecule; since the structure of the activated complex could not be predicted, it is simply postulated and the parameters applied. It is not the absolute values of these parameters which are usually meaningful, but their variation either in a series of not too different molecules or from one atom (bond) to another within the same molecule. These parameters point out the relative reactivity of molecules in a series and the sites in a molecule where substitutions, additions or other reactions occur, respectively.

Of the energy parameters, the resonance energy E_R is a global stability index, useful especially to discuss tautomeric equilibria in a series of molecules. The energy of the highest occupied orbital, ε_{HO}, indicates how easily the molecule is oxidised and donates electrons while the energy of the lowest empty orbital, ε_{LE}, indicates the ease with which the molecule is reduced and accepts electrons. The aromatic and stable systems have ε_{HO} and ε_{LE} much below and above $\varepsilon = \alpha$ (the energy characterising non-bonding orbitals) respectively. If $\varepsilon_{HO} > \alpha$ the molecule is very easily oxidised, tends to give a positive ion (or radical ion) while $\varepsilon_{LE} < \alpha$ points to an easily reduced molecule which gives a negative ion.

Among the electronic charge distribution parameters, the atomic charges, ρ_q, parallel the variation of basicity of the respective atom or its tendency to be alkylated (if the atom also possesses a lone pair of electrons), with the ease of electrophilic substitutions at the atom and antiparallels the ease of nucleophilic substitutions. Addition reactions of free atoms and radicals occur preferentially at atoms with largest free valence indexes, F_q.

Let us compare some of these indexes for molecules shown in Figure 2.4.

From the four molecules, the easiest to reduce should be acridine ($\varepsilon_{LE} = \alpha + 0.278\ \beta$) and to oxidise too ($\varepsilon_{HO} = \alpha + 0.657\ \beta$). Concerning the basicity of nitrogen atoms, N_9 of purine should be the most basic ($\rho = 1.308$) and that of pyrimidine the least basic ($\rho = 1.161$). The ease of electrophilic substitutions at carbon atom should increase in the order purine ($\rho_1 = 0.907$) < benzene ($\rho = 1.000$) < pyrimidine ($\rho_2 = 1.007$) < acridine ($\rho_4 = 1.016$). If fairly different molecules are compared, e.g. ethylene and benzene, the free valence indexes F_q (0.732 and 0.400 respectively) indicate that ethylene should give more readily additional reactions, which is actually the case, but both molecules should behave identically at oxidation, reduction, electrophilic and nucleophilic substitutions and should absorb in the same region of the ultraviolet spectrum—which is not the case. This illustrates the fact that these indices could not be used to compare molecules of too different structure. We also could not compare, for example, the basicity or tendency to radical addition of some different atoms (0 with N or C) on grounds of indices like ρ_q, F_q, etc.

I

$\varepsilon_{LE} = \alpha - 1.000\beta$

$\varepsilon_{HO} = \alpha + 1.000\beta$

II

$\varepsilon_{LE} = \alpha - 0.545\beta$

$\varepsilon_{HO} = \alpha + 0.687\beta$

III

$\varepsilon_{LE} = \alpha - 0.715\beta$

$\varepsilon_{HO} = \alpha + 0.774\beta$

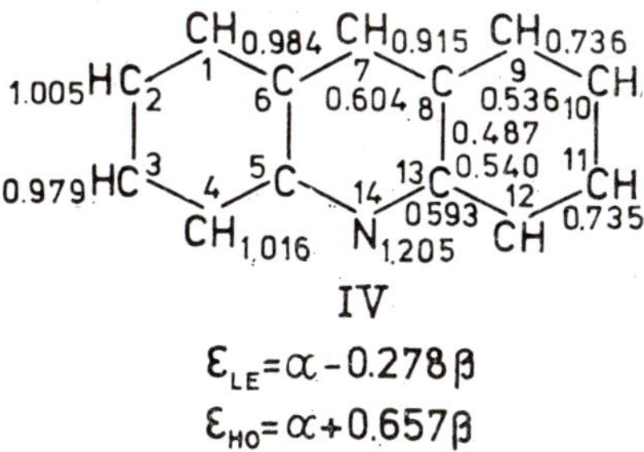

IV

$\varepsilon_{LE} = \alpha - 0.278\beta$

$\varepsilon_{HO} = \alpha + 0.657\beta$

Figure 2.4. π-electronic charges (figures next to atom symbols), bond orders (numerals next to bonds) for benzene (I), pyrimidine (II), purine (III) and acridine (IV) calculated by HMO method.

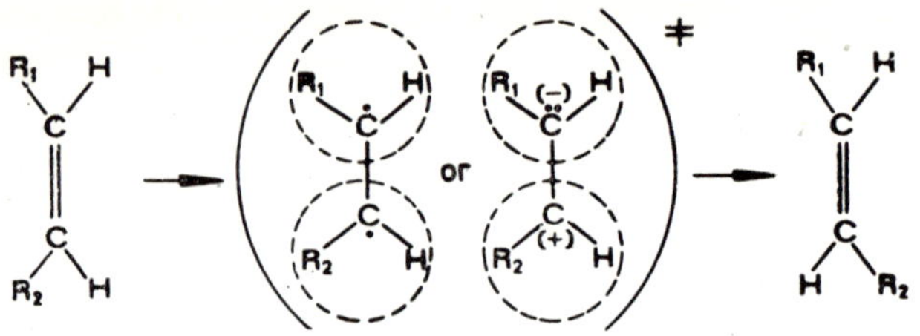

Figure **2.5.** Activated complex assumed for *cis-trans* isomerization reaction around the ethylenic double bond.

2.3.4. *Dynamic reactivity parameters*

These usually refer to calculation of the difference between the energy of the activated complex $E_{\pi}^{\neq}$, to which a certain structure is assigned, and that of the initial molecules, E_{π}^{0}. Thus for *cis — trans* isomerisations, the activation energy was calculated with good results by assimilating the activated complex with the molecule whose groups rotated 90° around the double bond with breaking of a π-bond:

$$E_{act} = E_{\pi}^{\neq} - E_{\pi}^{0} \qquad (2.10)$$

Two distinct conjugated subsystems are considered within the activated complex and the activation energy equals the energy increase caused by interruption of conjugation. The two conjugate subsystems have a radical character or anion and cation character if the difference between the two electronegativities, R_1 and R_2, is large enough [8].

Another parameter of this kind is the delocalisation energy, E_L. For a substitution at an atom of the conjugate system, two σ-bonds are considered at the atom, which is therefore left out of the conjugate system (together with 2π electrons in the case of electrophilic substitution). As an example for the electrophilic substitution in C_6H_6 one may offer:

$$E_L = E_{\pi}([\mathrm{RCH} \cdots \mathrm{CH} \cdots \mathrm{CH} \cdots \mathrm{CH} \cdots \mathrm{CHR}]^{+}) - E_{\pi}(C_6H_6) \qquad (2.10a)$$

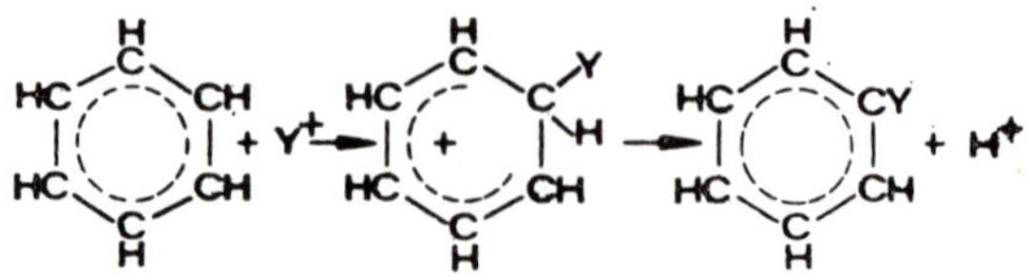

Figure **2.6.** Electrophilic substitution.

Comparison of E_L values for a substitution reaction at several not too different molecules or at atoms in a molecule, allows succesfully predicting of relative substitution rates. Table 2.2 lists (according to [1b], pp. 132, 141)

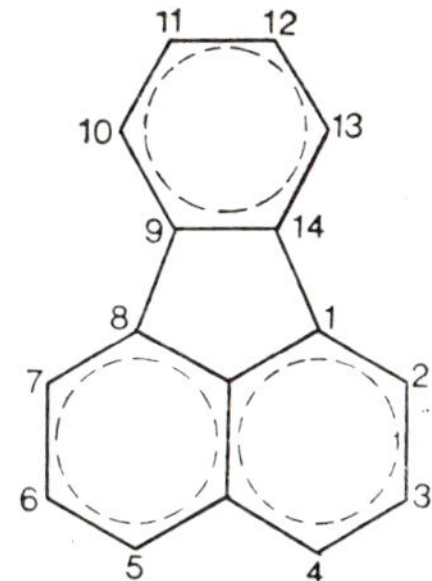

TABLE 2.2

Position	δ_q	F_q	$E_{L(q)}$
2	0.945	0.459	—1.77
3	1.005	0.398	—1.82
4	0.958	0.470	—1.67
10	0.997	0.438	—1.71
11	1.008	0.409	—1.76

the atomic charges, free valence indices and localisation energies for the electrophilic substitution of fluoranthene molecule. The electrophilic substitution should occur at C_{11} according to ρ_7 and at C_4 according to $E_{L(q)}$ (C_4 has the lowest E_L). The experiment show the substitution to take place at C_4, as predicted by E_L.

Fukui's frontier orbitals [2] may also be considered dynamic parameters that correspond to a perturbational approximation in localisation energy calculations. This index is based on the fact that, according to perturbational calculus, the substantial contribution to E_L is brought in either by the highest occupied molecular orbital if the molecule donates electrons in the activated complex (e. g. electrophilic substitution) or by the lowest empty orbital if the molecule accepts electrons (nucleophile substitution) or by both in radical addition reactions. The reactivity index of atom q, f_q, for substitution reactions in the respective molecule is given by [12]:

$$f_q = (2 - \nu)\, c^2_{q,\mathrm{HO}} + \nu c^2_{q,\mathrm{LE}} \tag{2.11}$$

with $\nu = 0$ for electrophilic substitutions, $\nu = 2$ for nucleophilic substitution and $\nu = 1$ for radical substitution. More intricate definitions for Fukui's reactivity indices than equation (2.11) were also given.

2.3.5. *The principle of conservation of orbital symmetry*

This was discussed by Woodward and Hoffman [9] and allows one to predict, to some extent, the structure of the activated complex of the reaction and whether or not the reaction requires a high activation energy.

This principle makes it possible to predict if some chemical bonds must be broken in the activated complex or, in case this is not so, if only a small activation energy may come out due to electrostatic repulsions or to a small steric repulsion occurring at the distance where new bonds begin to form. The method can inform us if a thermal or photochemical reaction may advance and if this is the case, which of the possible stereoisomers are formed.

The principle of this method consists of the following: during a chemical reaction the energies of molecular orbitals change uniformly and one can establish a correlation which indicates into what molecular orbital of the product(s) transforms each molecular orbital of the reactant(s). The molecular orbitals are arranged according to increasing energy ($\varepsilon_\sigma < \varepsilon_\pi < \varepsilon_n < \varepsilon_{\pi^*} < \varepsilon_{\sigma^*}$) and the correlation is established in terms of the principle of non-intersection of orbitals of identic al symmetry, with respect to symmetry elements which are preserved during the whole elementary reaction. The energy of reacting molecules is set equal to the sum of energies of electrons on occupied orbitals. If a bonding or non-bonding occupied molecular orbital of the reactant correlates with an antibonding molecular orbital of the products, an energy barrier arises which determines a high energy of activation for the reaction.

As an example, let us consider the addition of molecular and atomic hydrogen to ethylene. The symmetrical element which is conserved during these reactions is, in the first case, a symmetry plane perpendicular on the middle of molecules and, in the second case, a symmetry plane which contains the two C atoms, and the H atom so that all the orbitals involved are symmetric with respect to this element.

To illustrate these, here are the bonding and antibonding molecular orbitals of the ethylen e bond (Figure 2.7):

The bonding orbital of ethylenic π-bond is $\pi_{CC} = \pi_{C_1} + \pi_{C_2}$, more exactly 0.707 $(\varphi_1 + \varphi_2)$ and the antibonding one $\pi^*_{CC} = \pi_{C_1} - \pi_{C_2}$ and 0.707 $(\varphi_1 - \varphi_2)$ respectively. Herein $\pi_{C_1} = \varphi_1$ and $\pi_{C_2} = \varphi_2$ are the $2p_z$ atomic orbitals of the two C atoms. The two atomic orbitals, are for each other, the image on a mirror plane perpendicular on the middle of ethylene and hydrogen molecules. The bonding orbital, π_{CC}, rests unchanged, is symmetric with respect to the plane of symmetry defined above while the antibonding orbital, π^*_{CC}, changes only the sign and is antisymmetric.

In the case of the $C_2H_4 + H_2$ reaction, the bonding orbital π_{CC} of C_2H_4 correlates with the symmetric combination $\sigma^{*\prime}_{CH}$ of the two anti-

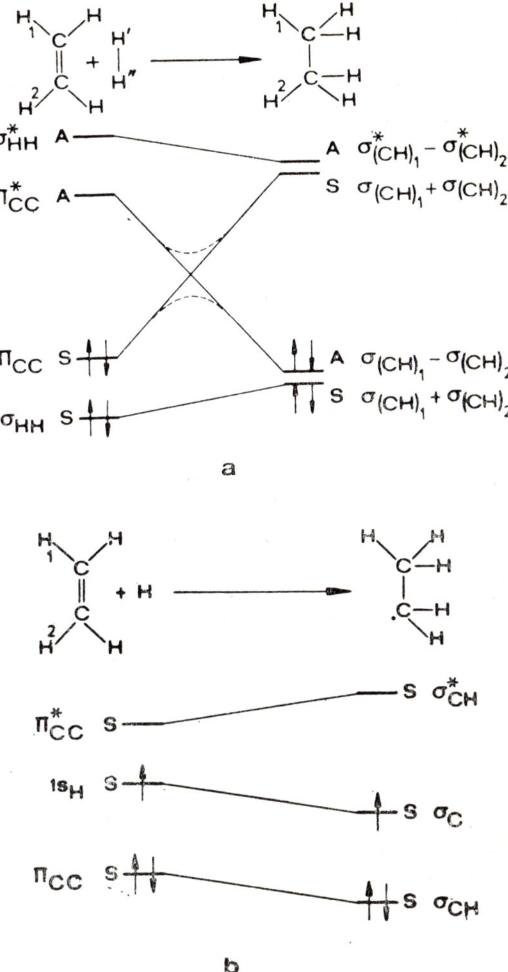

Figure 2.7. Woodward-Hoffman correlation diagram for: (a) reaction of ethylene with molecular hydrogen; (b) reaction of ethylene with atomic hydrogen.

bonding σ^*_{CH} orbitals newly formed in the product C_2H_6 so that the reaction is opposed to an energy barrier. The hydrogenation of ethylene does not take place through this molecular mechanism. In excited ethylene with an electron on π^*_{CC}, this antibonding orbital correlates with the asymmetric combination of σ'_{CH} bonding orbitals of the ethane molecule which makes the reaction possible. In the hydrogenation of ethylene with atomic H, the bonding π_{CC} correlates with σ_{CH} (newly formed), non-bonding $1s_H$ orbital with non-bonding σ_C orbital and antibonding π^*_{CC} with antibonding σ^*_{CH}. The hydrogenation of C_2H_4 with atomic H proceeds easily, the activation energy $E_{act} \cong 0$; the reverse reaction, the dissociation of an atomic H from the C_2H_5 radical requires a certain energy consumption since the energy of σ bonding orbital of the CH bond, σ_{CH} is lower than the energy of bonding π-orbital of CC bond, π_{CC}:

$$\varepsilon_\pi(CC) = \frac{\varepsilon_{C=C} - \varepsilon_{C-C}}{2} \cong -32 \text{ kcal mol}^{-1};$$

$$\varepsilon_\sigma(CH) = \frac{D_{C-H}}{2} \cong -49 \text{ kcal mol}^{-1}$$

This reasoning also holds for molecular (atomic) hydrogenation of unsymmetric substituted ethylene derivatives or for addition of a heteroatomic molecule (e.g. HCl). As the asymmetry plan perpendicular on the middle of bonds disappears, all orbitals become equivalent as symmetry and in their intersection (the activated complex) degeneracy is split (dotted line in Figure 2.7); however the tendency of π_{CC} orbital energy to increase during reaction persists and thus, although reduced, the potential barrier is still important.

2.4. **Hoffman molecular orbital method (EHT)** [10]

The version of the molecular orbital method issued by Hoffman, the so called extended Hückel theory (EHT) was especially conceived for the calculation of molecular energy as a function of the configuration of the nuclei, that is, for the evaluation of activation energies. It is also a uni-electronic method which does not account explicitly for interelectronic repulsions. Unlike the HMO method, it considers all unhybridised valence atom orbitals ($1s_H$, $2s$, $2p_x$, $2p_y$, $2p_z$ for elements of the second period). The S_{pq} overlap integrals are calculated at first as functions of internuclear distances, r_{pq}, (the x, y and z axes have fixed orientation in space). The hybridisation of atomic orbitals is not introduced *a priori* but results from the molecular orbitals.

The F_{pq} matrix elements:

$F_{qq} = -I_q$, valence state ionization potential corresponding to the atomic orbital (2.12)

$$F_{pq} = -0.875\ S_{pq}(r_{pq})\ (I_p - I_q) \quad (2.12a)$$

The energy and electron distribution parameters are similar to those of the HMO method; some discrepancies exist since for the overlapping of atomic orbitals one does not employ the zero differential overlap approximation. The charge densities ρ_q and bond orders p_{pq} contain terms in overlap integrals. The total energy is also considered to be a sum over occupied orbital energies (equation (2.5)), depending on interatomic distances r_{pq} plus the terms corresponding to electrostatic repulsion of nuclear cores. The repulsion between non-bonded atoms is for the most part due to the following effect: Wheland's version of the HMO method [2] (which considers the overlap between adjacent atoms to be non-zero) is characterised by a disymmetry in orbital energies in the sense that with respect to the mean value of atomic orbital energy $(1/m) \sum \alpha_q$, the antibonding molecular orbitals are raised more than the bonding ones lowered: $\sum_i \varepsilon_i = \sum_q \alpha_q$ in the HMO method but $\sum_i \varepsilon_i > \sum_q \alpha_q$ in the Wheland method. The left hand term of the inequality is greater, the larger the overlap of adjacent atoms. When the non-bonded atoms come closer, two bonding orbitals (each with two electrons) begin to interact and of the two resulting orbitals the energy of the "repulsive" one is more strongly raised than that of "attractive" one which is lowered.

The EHT method takes into account all valence electrons and atomic orbitals. It requires calculation of overlap integrals S_{pq} in terms of interatomic distances. If the activation energy of a reaction is to be calculated, the energy of reacting molecules (reaction complex) must be calculated for several sets of interatomic distances intermediary between reactant(s) and product(s).

The calculation of the activation energy of a reaction requires calculating the energy of reacting molecules (reacting complex) for several sets of interatomic separations ranging from those of reactant(s) to those of product(s) and locating the "saddle point" i.e. the top of the potential barrier. The EHT method is much more laborious than the HMO method and needs computers. However, its results are outstandingly interesting for chemists and generally agree fairly well with experimental data, at least for the relative activation energies in a class of reactions involving not too dissimilar molecules. The method was applied with good results to calculations in amides, esters, hydrogen bonded amides used as model combinations for peptides [10 a], etc.

Similar to the EHT method, the Wolfsberg-Helmholtz method [11] is useful in energy-level and charge-distribution calculations for transition metal complexes. Due to large differences in electronegativity (I_q), large electronic charge transfer from ligands to the central cation are obtained initially; therefore one uses an iterative procedure, the ionisation potential being considered to increase with the total (positive) atomic charge (s_{at}):

$$I_q = I_q^0 + as_{at} + \dots \quad s_{at} = Z_{at} - \sum_{q,\,\text{atom}} \rho_q \tag{2.13}$$

2.5. The Del Re method

The Del Re method was developed for the calculation of relative bond energies and atomic charges in systems of σ-bonds. The parameters employed in the calculations were adjusted so as to lead to good agreement between calculated and experimental dipole moment values in aliphatic compounds. It is basically a HMO method for dicentric molecular orbitals corresponding to localised σ-bonds. Approximate equations for calculation of bond charges, Q_{pq} (more exactly the charge displacement between the two atoms of the bond) and of bond energies, ε_{pq} are also obtained from this method:

$$Q_{pq} = \frac{\delta_p - \delta_q}{2\eta_{pq}} \tag{2.14}$$

$$\varepsilon_{pq} = 2\alpha + [\delta_p + \delta_q + \eta_{pq}\sqrt{1 + Q_{pq}^2}]\beta - E_{\text{rep}} \tag{2.14a}$$

with E_{rep} standing for the repulsion energies of positive atomic cores. The effective atomic charge ρ_q (in electron charge units with respect to electroneutrality) is calculated as the sum of bond charges at the given atom:

$$\rho_q = \sum_{p,\text{neighbours}} Q_{pq} \tag{2.14b}$$

The clue of this method is the calculation of δ_q parameters (which characterise the electronegativity of q atom orbitals) from the following system of equations which take into account the inductive effect upon atom q of neighbouring atoms bonded to it:

$$\delta_q = \delta_q^0 + \sum_{p,\ \text{neighbours}} \gamma_{q(p)} \cdot \delta_p; \quad q, p = 1, 2, \dots, N \tag{2.15}$$

where N is the number of atoms in molecule. The parameters required by Del Re calculations are accordingly δ_q^0 — proportional to standard electronegativity of the atom (δ_H^0 is considered zero), $\gamma_{q(p)}$ coefficients which introduce the inductive effect of atom p upon its neighbour q, and the exchange integrals η_{pq}. The $\gamma_{q(p)}$ values are selected according to semi-empirical reasoning and making use of experience with the inductive effect in π-electron systems where calculations give good results; η_{pq} values are basically proportional to the strength of σ_{pq} bond and practically are so selected as to lead to agreement of calculated and experimental dipole moments of some reference compounds. A list of recommended δ_q^0, $\gamma_{q(p)}$ and η_{pq} values is given in Table 2.3. For other atoms, δ_x^0 parameters can be calculated by Pullman's equation [13 a]:

$$\delta_x - \delta_c = \frac{3}{4}(\chi_x - \chi_c) - \frac{0.18}{2}(A_x - A_c)$$

where χ_x and χ_c are Mulliken's electronegativities, $\chi = 0.09 \cdot (I + A)$, A the electron affinity, I the ionisation potential in eV.

TABLE 2.3 **Parameters for Del Re method** [13, 13b].

$\delta_H^0 = 0.00$; $\delta^0_{C\,(aliphatic)} = 0.07$; $\delta^0_{-N\langle} = 0.24$; $\delta^0_{-O-} = 0.40$;

$\delta_F^0 = 0.57$; $\delta_{Cl}^0 = 0.35$

$\delta^0_{-C=} = 0.12$; $\delta^0_{-N=} = 0.38$; $\delta^0_{N\,(pyrollic)} = 0.30$; $\delta_{=O} = 0.28$

$\eta_{CH} = \eta_{CC} = 1.00$; $\eta_{CO} = 0.95$; $\eta_{NH} = \eta_{OH} = 0.45$; $\eta_{CCl} = 0.65$

$\eta_{C-N} = 0.7$; $\eta_{CO\,(carbonyl)} = 0.7$

$\gamma_{C(H)} = 0.3$; $\gamma_{C(C)} = \gamma_{C(N)} = \gamma_{C(O)} = \gamma_{C(F)} = 0.1$;

$\gamma_{N(H)} = \gamma_{O(H)} = 0.3$; $\gamma_{C(Cl)} = 0.2$

$\gamma_{H(C)} = 0.4$; $\gamma_{N(C)} = \gamma_{O(C)} = \gamma_{F(C)} = 0.1$; $\gamma_{H(N)} = \gamma_{H(O)} = \gamma_{H(Cl)} = 0.4$

This method yields for distribution of σ-electron charges good agreement with experimental data (calculated and measured dipole moments) and the energy parameters, ε_{pq}, correlate satisfactorily with some experimental magnitudes such as basicity of aliphatic amines. Combining the HMO method for π-electron systems with the Del Re method for σ-electrons and using suitably adjusted parameters, Pullman [20] has calculated the global atomic charge distributions ($\rho = \rho_\sigma + \rho_\pi$) for natural purine and pyrimidine bases as well as for synthetic analogues of these bases. The charge distributions so obtained agree well with those obtained by advanced methods which take into account all valence electrons. Sometimes the charge distributions calculated by HMO + Del Re methods give dipole moments which are in even better agreement with experiment, than those calculated with charge distributions resulting from advanced methods. Figure 2.8 lists some global atomic charges for natural purine and pyrimidine bases calculated by joint method [20]. For example, the calculated and experimental (in parantheses, only for part of the compounds) dipole moments are: guanine, 6.8 ; cytosine, 7.2 (8.0); thymine, 3.6; uracyl, (3.9); indole, 2.05 (2.05); imidazole, 4.03 (3.99); (all in Debyes).

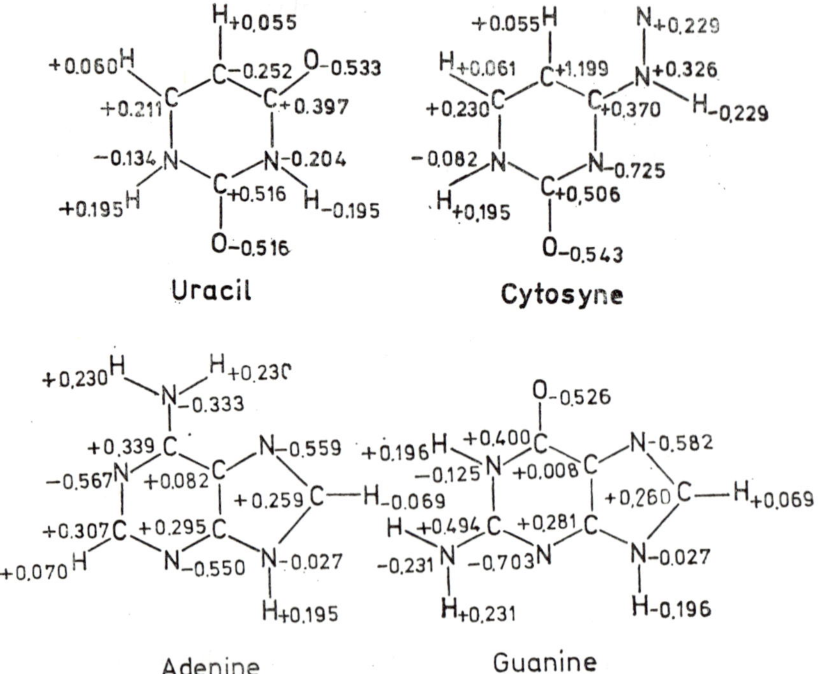

Figure **2.8.** Global atomic charges for nitrogen bases calculated by HMO + Del Re method.

2.6. **The Pariser-Parr-Pople method** [14]

The Pariser-Parr-Pople method is the simplest one which takes correctly into account the interelectronic repulsion and the asymmetry of wave function with respect to permutation of electrons. This is a self-consistent field method (SCF) corresponding to the general formalism developed for the molecular orbital methods by Roothaan [15]. The PPP method was developed for π-electron systems and also contains a series of parameters adjusted in accordance with some experimental data (atomic ionisation potentials, etc.). Unlike the methods discussed above, which do not consider the interelectronic repulsion and the asymmetry of wave function, the PPP method yields not only essentially correct charge distributions and molecular energy levels but equally correct electronic transition energies, accounting for the difference between excited electronic singlet and triplet states.

Calculations of the matrix elements F_{pq} of secular equations system (2.2), start up with the expression of the Hamiltonian for one electron. The energy and distribution of this electron depends on the attraction of atomic cores (atoms with π-electrons omitted) and the repulsion of other π-electrons. This repulsion is calculated by assuming from the start, a certain distribution for π-electrons. Solving the system of secular equations one ends up with a new distribution of π-electrons which is further employed in a more exact calculation of the effect of interelectronic repulsion on uni-electronic distribution. The procedure is re-iterated (iterative method) until the resulting electronic distribution is close enough to the previous distributions employed in calculating the interelectronic repulsion. The different integrals in F_{qq} and F_{pq} matrix elements are calculated by means of zero differential overlap approximation: all interelectronic repulsion integrals which contain $\varphi_p(r_i)$. $\varphi_q(r_i)$ products, i.e. products over atomic orbitals, functions of the same electronic coordinate r_i but centred at different atoms p and q, are neglected. The remaining integrals are either identified with empirical figures (ionisation integrals, I_q, of given atomic valence state) or calculated by approximations based on interelectronic coulombic repulsion of point charges. Finally the F_{qq} matrix elements for system (2.2) become:

$$F_{qq} = -I_q + \frac{1}{2}\rho_q \cdot \gamma_{qq} + \sum_{p \neq q} (\rho_p - Z_p)\,\gamma_{pq}$$

$$F_{pq} = \beta_{pq} - \frac{1}{2} p_{pq}\gamma_{pp}; \quad S_{pq} = \begin{cases} 1 \text{ for } p = q \\ 0 \text{ for } p \neq q \end{cases}. \tag{2.16}$$

The atomic π-electron charges, ρ_q and bond orders, p_{pq} are calculated according to equations (2.5a) and (2.5b). Z_p are the charges of atomic cores, equal to the number of π-electrons which the atom shares with the system, plus the formal charge of the atom. The exchange energy integrals β_{pq} are calculated by either an equation of type (2.12a) or considered proportional to β_{pq} integrals (recommended in HMO calculations) by simply assuming the standard $\beta \cong -2.5$ eV. γ_{qq} and γ_{pq} are energies of repulsion between two electrons, both on the q atomic orbital or one on p and one on q, respectively. They are calculated according to equations:

$$\gamma_{qq} = \frac{e^2}{r^0_{qq}} = I_q - A_q; \quad \gamma_{pq} = \frac{e^2}{r_{pq} + \frac{1}{2}(r^0_{qq} + r^0_{pp})}, \tag{2.16a}$$

with I_q and A_q representing the ionisation potential and electron affinity, respectively, of the corresponding valence state of atom q, r_{pq} the distance between nuclei of p and q atoms, while the standard distances r^0_q and r^0_p are calculated from the equation for γ_{qq}.

The energy of molecular orbital Φ_i becomes:

$$\varepsilon_i = \sum_q (c^2_{qi}F_{qq} + \sum_p c_{qi}c_{pi}F_{pq}) + \sum_{j,\ \text{occupied orbitals}} (2J_{ij} - K_{ij}) \tag{2.17}$$

with

$$J_{ij} = \sum_p\sum_q c^2_{ip}c^2_{jp}\gamma_{pq}; \quad K_{ij} = \sum_p\sum_q c_{ip}c_{iq}c_{jp}c_{jq}\gamma_{pq} \tag{2.17a}$$

The total energy for "closed shell" electronic configurations, i.e. with two electrons per orbital up to a limit, and none above it, is:

$$E = 2\sum_i\sum_q (c^2_{qi}F_{qq} + \sum_p c_{qi}c_{pi}F_{pq}) + \sum_i\sum_j (2J_{ij} - K_{ij}). \tag{2.17b}$$

The summation indexes i and j run only over the occupied molecular orbitals. The energy of transition from closed shell ground state to excited singlet state, originates in transition of one electron from the occupied molecular orbital, i, into the free one, l, and is given by:

$$\begin{aligned} {}^1E_{i\to l} = {}^1E(il) - {}^1E_0 = \varepsilon_l - \varepsilon_i + \sum_{j\text{-occupied}} (2J_{lj} - K_{jl}) \\ - \sum_{j\text{-occupied}} (2J_{ji} - K_{ji}) + J_{il} - J_{ii} + K_{il} \end{aligned} \tag{2.17c}$$

while for the same transition with an excited triplet state (S = 1) as outcome:

$$^3E_{i\to l} = {}^3E_{il} - {}^1E_0 = {}^1E_{il} - {}^1E_0 - 2K_{il}. \tag{2.17d}$$

For example, in case of ethylene there are two atomic $2p_z$ orbitals of C atoms and 2π-electrons. Because of symmetry reasons, the molecular orbitals are:

$$\begin{aligned} \Phi_2 &= 0.707\,(\varphi_1 - \varphi_2) \\ \Phi_1 &= 0.707\,(\varphi_1 + \varphi_2) \end{aligned} \tag{2.18}$$

with two electrons on Φ_1, in ground state. The energies of the two orbitals are:

$$\begin{aligned} \varepsilon_2 &= F_{CC} - 0.5\, F_{CC'} + 2J_{21} - K_{21} \\ \varepsilon_1 &= F_{CC} + 0.5\, F_{CC'} + 2J_{11} - K_{11} \end{aligned} \tag{2.18a}$$

with

$$F_{CC} = - I_C + 0.5\,\gamma_{CC} + (1.00 - 1)\,\gamma_{CC'}$$

$$F_{CC'} = \beta_{CC} - 0.5\,\gamma_{CC'}$$

$$J_{11} = J_{21} = 0.25(2\,\gamma_{CC} + 2\,\gamma_{CC'}); \quad K_{11} = 0.25(2\,\gamma_{CC} + 2\,\gamma_{CC'});$$

$$K_{12} = 0.25(2\,\gamma_{CC} - 2\,\gamma_{CC'}) \tag{2.18b}$$

$$\gamma_{CC} = \frac{e^2}{r^0_{CC}} = I_C - A_C; \quad \gamma_{CC'} = \frac{e^2}{r^0_{CC} + r_{CC'}}$$

$$^1E_{1\to 2} = \varepsilon_2 - \varepsilon_1 + J_{12} - J_{11}; \quad ^3E_{1\to 2} = \varepsilon_2 - \varepsilon_1 + J_{12} - J_{11} - 2\,K_{12}$$

Concerning parameter selection of the PPP method, a review on ionisation potentials and electron affinities of various atomic valence states was given by Hinze and Jaffé [16]. The β_{xy} integrals are often chosen so as to ensure the agreement between calculated and experimental transition energies of the first term in the series of molecules with heteroatoms under study. The interatomic distances r_{pl} are calculated from the spatial structure of the molecule as given by various methods (X-ray diffraction, etc.). Some of the most useful parameters are listed in Table (2.4).

The PPP method was applied most successfully to interpretation of the electronic spectra of organic molecules. In this case too, the parameters were established mostly for atoms of the second period, and to a smaller extent for atoms of following periods. The best results of *nd* orbital calculations for pentavalent tetracoordinated P atoms, were obtained

TABLE 2.4 **Parameters for PPP method [16]**

Atom	*Valence state*	*I* (eV)	*A* (eV)
C	(di di π)π	11.19	0.10
	(tr tr tr)π	11.16	0.03
	σ	15.62	1.95
	(te te te)te	14.61	1.34
N	(tr tr tr)π	14.12	1.78
	σ	20.60	1.78
	(tr tr tr)π^2	19.72	4.92
	(te te te)te	18.93	4.15
O	(di di π^2)π^2	28.71	9.51
	(tr^2 tr^2 tr)π	17.70	2.47
	σ	26.65	7.49
	(tr^2 tr tr)π^2	26.14	7.32
	(te^2 te^2 te)te	24.39	6.11
F	(s^2 p^2 p^2)p	20.86	3.50
Cl	(s^2 p^2 p^2)p	15.03	3.73
B	(tr^0 tr tr) [17]	8.33	
P	(tr^2 tr tr)π	11.64	
	(tr^2 tr tr)dπ	2.91	
	(te^2 te te)te	5.79	
	free atom	10.98	0.77
S	free atom	10.36	2.07

di-digonal hybridisation

tr-trigonal hybridisation

te-tetragonal hybridisation

σ, π — refer to σ and π-orbitals

Exchange energy integrals [17]

$$\beta^{pp}_{XY} = -0.45\left(I^p_X + I^p_Y\right) \cdot S^{pp}_{XY}$$

$$\beta^{dp}_{XY} = -0.39\left(I^d_X + I^p_X\right) S^{dp}_{XY}$$

with two $3d$ orbitals orthogonal to P and one π-electron [5]; the best parameters compared to data from reference [17] seem to be:

$$I_{\mathrm{P}}^{3d} = -2.9 \text{ eV}; \quad \gamma_{\mathrm{PP}}^{3d} = \gamma_{\mathrm{P'P}}^{3d} = 2.1 \text{ eV}; \quad \beta_{\mathrm{CP}}^{pd} =$$

$$= -1.0 \text{ eV } (S_{\mathrm{CP}}^{pd} \cong 0.20)$$

Table 2.5 lists the results of some PPP calculations of ionisation potentials (the energy of highest occupied orbital) and wave length of first absorption maximum in the electron spectrum (the energy $E(^1\mathrm{F} \rightarrow {}^1\mathrm{I})$ of the first singlet-singlet transition) for a series of nitrogen bases, as given by Pullman [20].

TABLE 2.5 **Ionisation potentials and transition energies for nitrogen bases**

Compound	$-\varepsilon_{\mathrm{HO}}$(eV)	I(eV) (exp)	$E(^1\mathrm{F} \rightarrow {}^1\mathrm{I})$ (eV)	λ_{max} exp	
				eV	nm
Guanine	7.59	7.8—8	4.4	4.5	275
Adenine	7.92	8.0—8.2	5.2	4.8	260
Hypoxantine	8.00		4.4	4.4	278
Cytosine	8.16	8.1—8.4	4.2	4.5	277
Purine	8.87		5.5	4.7	265
Uric acid	8.66		4.6	4.3	286
Xantine	8.82		5.0	4.6	267
Uracyl	9.15		5.1	4.8	258

2.7. **Advanced methods** [18]

The development of more powerful computers has oriented quantum chemistry in recent years, towards more exact calculations, with the elimination of simplifying approximations. The dividing of electrons into σ and π systems was abandoned and all valence electrons were taken into account. One starts with linear combinations of atomic orbitals as molecular orbitals and one applies the iterative Roothaan formalism [15]. There are several versions of these, the so called "all valence electrons - SCF" methods. Unlike the PPP method wich identifies integrals over atomic orbitals with different experimental data, the advanced methods calculate effectively at least part of these integrals, helped by the atomic orbitals, usually of the Slater type.

In these methods, the energy is written as

$$E = \frac{\int \psi^* \hat{H} \psi \, dv}{\int \psi^* \psi \, dv} \tag{2.19}$$

where ψ is an electronic function antisymmetrized with respect to permutation of electrons over molecular spinorbitals of (2a) type. The expression of energy is expanded into integrals over atomic orbitals. The general expression of Hamiltonian is:

$$\hat{H} = \sum_{\nu} (T_\nu + U_\nu) + \sum_{\nu, \mu} \frac{e^2}{r_{\nu\mu}} \tag{2.19a}$$

where T_ν denotes kinetic energy of electron ν, U_ν the interaction energy of electron ν with all atomic residues and $e^2/r_{\nu\mu}$ the repulsion energy between electron ν and μ. The energy E is thus a function of c_{kj} coefficients of molecular orbitals (equation 2.1) and of integrals:

monocentric:

$$\alpha_q \equiv \int \varphi_q(r_\nu)\,(T_\nu + U_\nu)\,\varphi_q(r_\nu)\,\mathrm{d}V_\nu \quad \text{and} \quad \int \varphi_q(r_\nu)^2 \frac{e^2}{r_{\nu\mu}} \varphi_q(r_\mu)^2 \mathrm{d}V_\mu\,\mathrm{d}V_\nu,$$

bicentric: $\beta_{pq} \equiv \int \varphi_q(r_\nu)\,(T_\nu + U_\nu)\,\varphi_p(r_\nu)\,\mathrm{d}V_\nu$ and

$$\int \varphi_q^*(r_\nu)\,\varphi_p(r_\nu)\,\frac{e^2}{r}\,\varphi_q^*(r_\mu)\,\varphi_p(r_\mu)\,\mathrm{d}V_\mu\,\mathrm{d}V_\nu, \tag{2.19 b}$$

tricentric: $\int \varphi_q^*(r_\nu)\,\varphi_p(r_\nu)\,\frac{e^2}{r_{\mu\nu}}\,\varphi_l^*(r_\mu)\,\varphi_l\,(r_\mu)\,\mathrm{d}V_\mu\,\mathrm{d}V_\nu,$

tetracentric: $\int \varphi_q^*(r_\nu)\,\varphi_p(r_\nu)\,\frac{e^2}{r_{\nu\mu}}\,\varphi_l(r_\mu)\,\varphi_k(r_\mu)\,\mathrm{d}V_\mu\,\mathrm{d}V_\nu.$

The number of integrals involved increases strongly in the sequence monocentric < bicentric < tricentric < tetracentric; e. g. for C_2H_6, with a total of 16 atomic orbitals there are 246 monocentric integrals, 2680 bicentric and 6390 tri- and tetracentric ones. On the other hand, the higher the polycentricity of an integral, the more products between the number of orbitals centred at different atoms and the smaller it is. In order that calculations do not get too laborious even for existing computers, the integrals of higher polycentricity are given up, if molecules are not too small. According to the way they consider polycentric integrals, the advanced methods may be classified as follows:

CNDO methods — complete neglect of differential overlap and thus all polycentric integrals containing products like $\varphi_q(r_\nu)$ $\varphi_p(r_\nu)$ except $\beta_{\mu\nu}$ integrals; the PPP method is a CNDO method.

MINDO methods — modified intermediate neglect of differential overlap, takes into account differential overlap of orbitals centred at the same atom but only for bicentric repulsion integrals.

PNDO methods — partial neglect of differential overlap with different variants. Usually integrals of overlap between orbitals centred at different atoms are neglected.

These methods may exist in different variants, depending on how the integrals not neglected are calculated or identified with experimental magnitudes (as ionisation potentials, empirical constanst of atomic spectra related to repulsion integrals for orbitals centred on the atom, etc.).

These methods have been applied to several inorganic molecules as H_2O, NH_3, HCl and organic ones — CH_4, H_2CO, benzene, naphthalene, etc., generally molecules with not much over 20 atoms. It is a common feature of these methods that none of them leads to results in agreement with all experimental data. The CNDO/2 method, for instance, gives good agreement for bond angles, dipole moments, force constants and bond lengths but rather bad results for heats of formation, energies of ionisation and spectral transitions. The MINDO/2 method applied to hydrocarbons gives good results for heats of formation and molecular geometry but does not agree with spectral data. These advanced methods give, generally, better results, the less extensive the neglecting of polycentric integrals and the more exact the evaluation of terms resulted from development over atomic centres i.e. of α_q and β_{pq} integrals; meanwhile computations are more laborious and extend only to atoms with fewer electrons.

Obviously, wherever computing methods, size and number of molecules of interest make it possible, the application of advanced methods described in this paragraph are preferable to simpler ones.

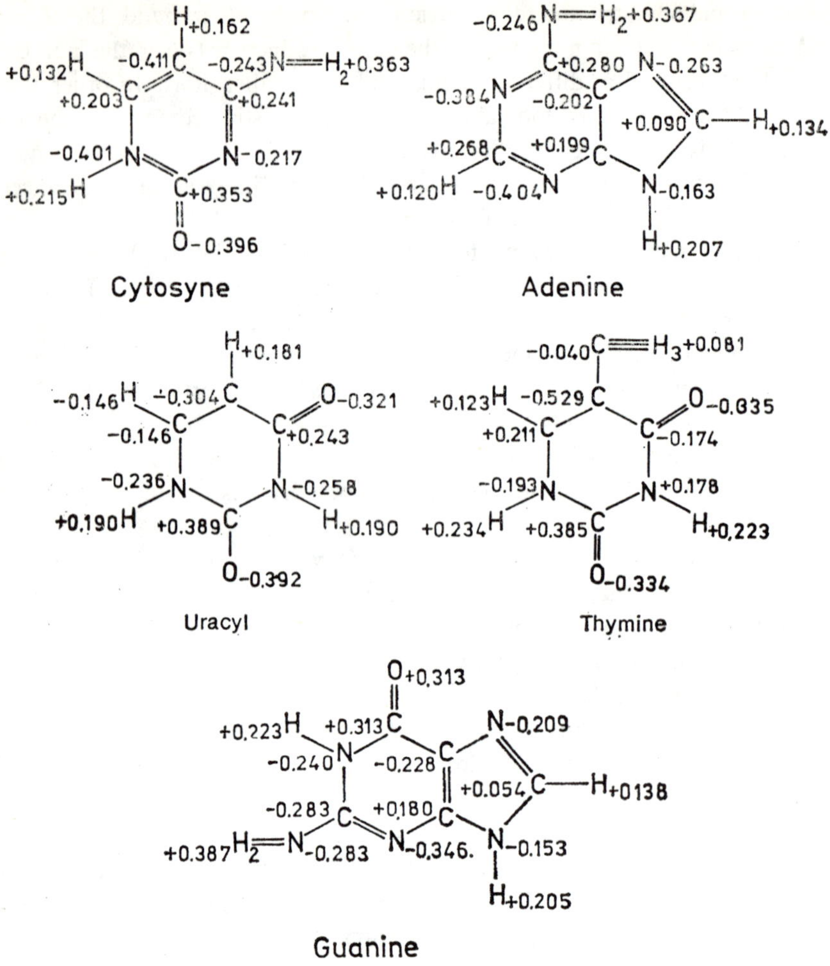

Cytosyne

Adenine

Uracyl

Thymine

Guanine

Figure 2.9. Total atomic charges for natural nitrogen bases calculated by AVE—CI—SCF method [25].

Numerous HMO calculations have dealt with nitrogen bases, especially those of Pullman's group, whose book [1] lists, in an appendix, molecular orbital energies and molecular diagrams of more than 80 purinic and purimidinic bases, their different tautomers and derivatives. From the more advanced calculations we remark upon Fraga and Carbo's 21] by "AVE—CI—SCF" method, which makes use of antisymmetrised self consistent field molecular orbitals with atomic orbitals whose electronegativity varies with total atomic charge; the method takes into account the configuration interaction and all valence electrons. The results for 83 bases are: total energies of ground states and first excited singlet and triplet states, energy and strength of oscillator for electronic transitions, total atomic charges for base, negative and positive radical ions and dipole moments. Figure 2.9 lists the total atomic charges calculated by Fraga and Carbo [21] for five natural bases.

A direct comparison of charge distributions in Figures 2.9 and 2.3 (the latter given by the Hückel method, for π-electrons only) could not be made. However, it is evident that the polarisation effect of σ-bonds cannot be neglected; the signs of effective charges of aminic and pyrolic nitrogen atoms, calculated by the two methods are different. A general parallel is obtained between the π-electron charge of an atom at different positions in the same or different molecules and Fraga and Carbo's total charge. The agreement between charge distributions obtained by Fraga and Carbo and by HMO + Del Re methods is rather good. There is a characteristic discrepancy: the HMO + Del Re methods frequently leads to absolute values of atomic charges considerably higher than the advanced method, since the latter takes into consideration more exactly the interelectronic repulsion, which opposes the clustering of electron sat one given atom.

REFERENCES

1. B. Pullman and A. Pullman, *Quantum Biochemistry*, John Wiley, Interscience New York (1963), Ch. 1 and 2; 1a: Ch. 3; 1b: Ch. 4.
2. A. Streitwieser jr., *Molecular Orbital Theory for Organic Chemists*, John Wiley, New York, (1961); 2a. Cl. Nicolau and Z. Simon, *Molecular Biophysics* (in Romanian), Scientific Publishing House, Bucharest, (1968) Ch. III.
3. W. R. Purcell and J. A. Singer, *J. Chem. and Eng. Data*, (1967) **12**, 235.
4. M. J. Dewar, E. A. Lucken and M. A. Whitehead, *J. Chem. Soc.*, (1960) 2423.
5. R. Vîlceanu, A. Balint and Z. Simon, *Nature*, (1968) **217**, 61; M. Mracec and Z. Simon, *Rev. Roum. Chim.*, (1971) **16**, 449; R. Vîlceanu, Z. Simon, A. Balint and M. Mracec, in *Chimie Organique du Phosphore*, Edit. CNRS, Paris, (1970), p. 39.
6. C. Sandorfy, *Canad. J. Chem.*, **33**, 1337 (1955); K. Fukui, H. Kato, T. Yonezawa *et al.*, *Bull. Chem. Soc. Japan*, **33**, 1197, 1201 (1950); **35**, 38, 1814 (1962).
7. M. Kasha, *Ultraviolet Radiation Effects. Molecular Photochemistry. Comparative Effects of Radiation*, John Wiley, New York; Z. Simon, *Studii Cercet. Chimie*, **13**, 697 (1964).
8. J. C. Murgulescu and Z. Simon, *Zt. Phys. Chem.* (Leipzig), **221**, 29 (1962).

9. R. B. Woodward and R. Hoffman, *The Conservation of Orbital Symmetry*, Chemie Verlag, Weinheim, (1970).
10. R. Hoffman, *J. Chem. Phys.*, **391,** 397 (1963); **40,** 2047, 2474, 2480 (1964); *J. Amer. Chem. Soc.*, **86,** 1259 (1964); *Tetrahedron*, **22,** 539 (1966); (a) J. F. Yan, F. A. Momany, R. Hoffman and H. A. Scheraga, *J. Phys. Chem.*, **74,** 2424 (1970).
11. M. Wolfsberg and L. Helmholtz, *J. Chem. Phys.*, **20,** 837 (1952).
12. K. Fukui, T. Yonezawa and Ch. Nagata, *Bull. Chem. Soc. Japan*, **27,** 423 (1954), *J. Chem. Phys.*, **26,** 831 (1957).
13. G. Del Re, *J. Chem. Soc.*, 4031 (1958); G. Del Re and B. Pullman, *J. Chem. Phys.*, **62,** 942 (1965); (a) A. Pullman, *Modern Quantum Chemistry*, **3,** 283 (1965); (b) H. Berthod and A. Pullman, *J. Chem. Phys.*, **62,** 942 (1965).
14. R. Pariser and R. G. Parr, *J. Chem. Phys.*, **21,** 466, 767 (1953); J. A. Pople, *Trans. Faraday Soc.*, **49,** 1375 (1953); K. Nishimoto and N. Mataga, *Z. Physik, Chem.*, **12,** 335 (1957).
15. C. J. Roothaan, *Rev. Mod. Phys.*, **23,** 69 (1951).
16. J. Hinze and H. H. Jaffé, **84,** 540 (1962); H. O. Pritchard and H. A. Skines, *Chem. Rev.*, **55,** (1955).
17. D. A. Brown and C. G. Bormack, *J. Chem. Soc.*, 5385, (1964).
18. G. Klopman and B. O'Leary, in *All-Valence-Electrons S. C. F. Calculations, of Large Organic Molecules*, Springer Verlag, Berlin (1970).
19. L. N. Ferguson, in *The Modern Structural Theory of Organic Chemistry*, Prentice Hall, New Jersey, p. 179 (1964).
20. B. Pullman, *Advan. Quantum Chem.*, **4,** 267, (1968).
21. S. Fraga and R. Carbo, *Technical Report TC-6914* (1969), University of Alberta.

3. Intermolecular Forces

Intermolecular forces are of first importance in biology; many interactions, like repressor-operator and antigen-antibody, are based upon these forces, without the formation of an actual chemical bond. Intermolecular forces between various biomolecules and between biomolecules and the surrounding aqueous solution also determine the spatial configuration of biomolecules. It is also these forces which determine the formation of enzyme-substrate complexes, the active transport, the partition of micromolecules among various phases (in the intracellular liquid or absorbed on membraneous formations). It is likely that intermolecular forces play a major part in the assembling of subcellular organites (polyribosomes, mitochondria), although formation of disulphide bridges may have its own share.

When dealing with intermolecular forces, two situations are observed: either direct interactions between two biomolecules, or two parts of them in which case the partners have a fixed spatial configuration, or interactions between biomolecules and aqueous solution (within which the configuration of water molecules and of ions do change in time) and interactions between two molecules through a layer of one or several water molecules. Evidently the calculation of interaction energy is more difficult in the second situation.

The calculation of interaction energy (at least in the first situation mentioned above) may be carried out rigorously by means of perturbation calculus starting with approximate wave functions and approximate energies for the two partners in their ground and first excited electronic states. From this, accurate, though hard to apply, scheme approximations may be deduced which correspond to different degrees of simplification and whose parameters are standardised on the basis of experimental data. Simplification carried through to the end allows a classification of intermolecular forces on empirical grounds: interactions between ions (in aqueous solutions of high dielectric constant so that "ionic bond" is as strong as other intermolecular forces), ion-dipole interactions, dipole-dipole interactions, van der Waals forces, repulsive forces, H—bonds, charge—transfer interactions, hydrophobic forces (representing the overall effect of several kinds of forces). On the basis of experimental data—association and partition equilibria—approximate values may be assigned to the various kinds of forces at the frontier of any two groups, from the above classification.

3.1. Semiempirical approaches of intermolecular forces

The interaction energy U between two molecules, A and B may be conveniently expressed as a first-order and second-order approximation of perturbation theory ($\Delta E^{(1)}$ and $\Delta E^{(2)}$ respectively):

$$U = \Delta E^{(1)} + \Delta E^{(2)} \therefore \Delta E^{(1)} = \langle \psi_0 | \hat{V} | \psi_0 \rangle;$$

$$\Delta E^{(2)} = \sum_e \frac{|\langle \psi_0 | \hat{V} | \psi_e \rangle|^2}{E_0 - E_e} \tag{3.1}$$

$\Delta E^{(1)}$ term corresponds to ordinary electrostatic interaction while $\Delta E^{(2}$ corresponds to the polarisation or dispersion term as the excited states belong to one molecule or to both molecules respectively. Function ψ belongs to the combined system, AB, subscript "0" to ground state and "e" to excited state. The interaction operator $\hat{V}$ has the expression:

$$\hat{V} = \sum_{\alpha=1,}^{M} \sum_{\beta'=1}^{M'} \frac{Z_\alpha Z_{\beta'} e^2}{R_{\alpha\beta'}} - \left(\sum_{\alpha=1,}^{M} \sum_{j'=1}^{N'} \frac{Z_\alpha e^2}{R_{\alpha j'}} + \sum_{\beta'=1,}^{M'} \sum_{i=1}^{N} \frac{Z_{\beta'} e^2}{R_{\beta' i}} \right) + \\ + \sum_{i=1,}^{N} \sum_{j'=1}^{N'} \frac{e^2}{R_{ij'}} . \tag{3.2}$$

The unprimed indexes, α and i, denote nuclei and electrons of molecule A, while the primed indexes, β' and j', the same for B. Let molecule A have M atomic cores (nuclei) and N electrons, and let M' and N' have similar meanings for B. $R_{\alpha\beta'}$, $R_{\alpha j'}$, $R_{ij'}$ represent the nucleus-nucleus, nucleus-electron and electron-electron distances between particles, one for each molecule, Z_α, Z_β the charges of nuclei (atomic cores). If the wave function is expressed as products of molecular orbitals, and these in their turn, as linear combinations of atomic orbitals and if one introduces the zero differential overlap approximation for atomic orbitals, then the three energy terms of $\hat{V}$ operator become ($U = E_{el} + E_{pol} + E_{dis}$):

$$E_{el} = \sum_{\alpha=1,}^{M} \sum_{\beta'=1}^{M'} \frac{\rho_\alpha \rho_{\beta'} e^2}{R_{\alpha\beta'}} \tag{3.3a}$$

$$E_{pol} = 2\sum_{s=1,}^{n} \sum_{l=n+1}^{\infty} \left\{ \frac{\left(\sum_{\beta'=1,}^{M'} \sum_{\mu=1}^{m} c_{s\mu} c_{l\mu} \frac{\rho_{\beta'} e^2}{R_{\beta'\mu}} \right)^2}{E_0^A - E_{s\to l}^B} \right\} + 2\sum_{k=1,}^{n'} \sum_{p=n'+1}^{\infty} \left\{ \frac{\left(\sum_{\alpha=1,}^{M} \sum_{\mu'=1}^{m'} c_{k\mu'} c_{p\mu'} \frac{\rho_\alpha e^2}{R_{\alpha\mu'}} \right)^2}{E_0^B - E_{k\to p}^A} \right\} \tag{3.3b}$$

$$E_{dis} = 4\sum_{s=1,}^{n} \sum_{l=n+1}^{\infty} \sum_{k=1,}^{n'} \sum_{p=n'+1}^{\infty} \left\{ \frac{\left(\sum_{\mu=1,}^{m} \sum_{\mu'=1}^{m'} \frac{c_{s\mu} c_{l\mu} c_{k\mu'} c_{p\mu'} e^2}{R_{\mu\mu'}} \right)^2}{E_0^A - E_{s\to l}^A + E_0^B - E_{k\to p}^B} \right\} \tag{3.3c}$$

where ρ_α is the net charge of atomic centre α; $c_{s\mu}$ — the coefficient of μ-th atomic orbital of molecular orbital s; m — the number of atomic orbitals (on molecule A), $E_{s\to l}^A$ the energy of an excited state corresponding to transition of an electron from orbital s on orbital l; E_0^A — the energy of ground state; n — number of occupied molecular orbitals, all A; the other symbols refer to B.

In a semiclassic interpretation, E_{el} (equation (3.3a)) represents the interaction between static charges of the two molecules, E_{pol} — the interaction of static charges of one molecule with those they induce in another molecule and the reverse, while E_{dis} represents the interaction between induced charges. The effective calculation of E_{pol} and E_{dis} by equations (3.3b) and (3.3c) requires the inclusion of a large number of excited states and was performed only for relatively small molecules [1]. Various approximations originating in these equations have been used for molecules of biologic interest.

3.1.1. *Bradley's monopole approximation — nitrogen bases* [3, 4]

This approximation is referred to as "monopole" since it performes calculations on the basis of interactions between static charges and (or) charges induced in atomic centres located each on a molecule. For calculations of induced charges, this method does not resort to perturbational treatment but considers the polarisation of different atoms or bonds of a molecule produced by the field of electrostatic charges on the other molecule and makes use of polarisabilities (from experimental data) of atoms (bonds) of the first molecule. The equations which define the three types of energies written in tensorial form are [4]:

$$E_{el} = \sum_{\alpha=1,}^{M} \sum_{\beta'=1}^{M'} \frac{\rho_\alpha \rho_\beta e^2}{R_{\alpha\beta'}} \tag{3.4a}$$

$$E_{pol} = -\frac{1}{2} \sum_{\alpha=1,}^{M} \sum_{\beta'=1}^{M'} \vec{\mathcal{E}}_{\alpha\beta'} [A_{\beta'}] \vec{\mathcal{E}}_{\alpha\beta'} - \frac{1}{2} \sum_{\beta'=1}^{M'} \sum_{\alpha=1}^{M} \vec{\mathcal{E}}_{\beta'\alpha} [A_\alpha] \vec{\mathcal{E}}_{\beta'\alpha} \tag{3.4b}$$

$$E_{dis} = -\frac{1}{4} \frac{I_A I_B}{I_A + I_B} \sum_{\alpha=1,}^{M} \sum_{\beta'=1}^{M'} \frac{1}{R^6_{\alpha\beta'}} \cdot Tr([T_{\beta'\alpha}][A_\alpha][T_{\alpha\beta'}][A_{\beta'}]). \tag{3.4c}$$

The field produced by the static charge of atomic centre α in the point where the atomic centre β lies, is:

$$\vec{\mathcal{E}}_{\alpha\beta'} = \frac{e\rho_\alpha}{R^3_{\alpha\beta'}} \cdot \vec{R}_{\alpha\beta'}$$

when $[A_\alpha]$ is the tensor of polarisability components for atom α, $[T_{\alpha\beta}]$ a tensor connected to $\vec{R}_{\alpha\beta'}$, the distance between atomic centres α and β′. *Tr* is the spur (sum of diagonal elements) of the matrix corresponding to the product of the tensors. The two tensors have the following expressions:

$$[A] = \begin{vmatrix} \alpha_{xx} & \alpha_{xy} & \alpha_{xz} \\ \alpha_{yx} & \alpha_{yy} & \alpha_{yz} \\ \alpha_{zx} & \alpha_{zy} & \alpha_{zz} \end{vmatrix};$$

$$[T] = \frac{3}{R^2} \cdot \begin{vmatrix} R_x^2 & R_x R_y & R_x R_z \\ R_y R_x & R_y^2 & R_y R_z \\ R_z R_x & R_z R_y & R_z^2 \end{vmatrix} - \begin{vmatrix} 1 & 0 & 0 \\ 0 & 1 & 0 \\ 0 & 0 & 1 \end{vmatrix}.$$

The components of polarisabilities α_{xx}, α_{xy}, etc. are calculated from atomic polarisabilities which are supplied by molecular refraction measurements. Frequently, only the diagonal terms are considered: $\alpha_{xx} = \alpha_{yy} = \alpha_\perp$, the polarisability in to the plane of the molecule and $\alpha_{zz} = \alpha_{||}$ the polarisability perpendicular the plane of the molecule. I_A and I_B represent the ionisation potentials of molecules A and B, respectively. The overall atomic charges, ρ_α, are calculated by one of the methods described in Chapter 2. For example, for nitrogen bases, $\rho_{\alpha'}$ values may be considered a sum of π-electron charges (calculated by HMO or PPP methods) and σ-electron ones (calculated by the Del Re method) [4]. The $\vec{R}_{\alpha\beta'}$ distances are estimated from the known geometry of interacting molecules.

If we focus on short-range repulsion between non-bonded atoms, a repulsion term, E_{rep}, varying with distance as $1/R^{12}$ must be included into the expression of interaction energy (i.e. $U = E_{el} + E_{pol} + E_{dis} + E_{rep}$). Thus according to Bradley [5] the interaction energy may be written as the sum of interatomic interaction potentials, $u_{\alpha\beta'}$ of the form:

$$E = \sum_{\alpha=1,}^{M} \sum_{\beta'=1}^{M'} u_{\alpha\beta'} \therefore$$

$$u_{\alpha\beta'} = A_{\alpha\beta'} R_{\alpha\beta'}^{-1} + B_{\alpha\beta'} R_{\alpha\beta'}^{-4} + C_{\alpha\beta'} R_{\alpha\beta'}^{-6} + D_{\alpha\beta'} R_{\alpha\beta'}^{-12}, \qquad (3.5)$$

where A, B, C, D are parameters characteristic to each pair of atoms. It is worth mentioning that equations (3.4a-c) are derived from equations (3.3a-c) by replacing the perturbational terms of atomic (bond) polarisabilities with experimental values. These relations correspond to the approximation of zero differential overlap between atomic orbitals and therefore do not account for the repulsion associated with the overlap of orbitals (occupied with electrons) from non-bonded atoms. Generally, the interaction between non-bonded atoms, $E_{dis}(\alpha\beta') + E_{rep}(\alpha\beta')$, may be described by a so called 6—12 Lenard-Jones potential (the last two terms in equation (3.5)), or a Buckingham potential:

$$E_{dis}(\alpha\beta') + E_{rep}(\alpha\beta') = -C_{\alpha\beta'} R_{\alpha\beta'}^{-6} + D_{\alpha\beta'} e^{-b_{\alpha\beta'} R_{\alpha\beta'}} \qquad (3.6)$$

In Table 3.1, C, D and b coefficients according to Liquori [6], for interactions between atoms are listed; the energies are in kcal mol^{-1} and distances in Å.

TABLE 3.1 **Coefficients of interaction between non-bonded atoms**

Atoms $\alpha\beta'$	$C_{\alpha\beta'}$	$D_{\alpha\beta'} \times 10^{-3}$	$b_{\alpha\beta'}$
C— — — C	298	237	4.32
C— — — O	244	212	4,44
C— — — N	244	212	4.44
C— — — H	121	31	4.20
O— — — O	200	186	4.55
O— — — N	200	186	4.55
O— — — H	99	28	4.32
N— — — N	200	186	4.55
N— — — H	99	28	4.32
H— — — H	49	6.6	4.08

Calculations of energies of interaction between nitrogen bases were carried out by Pullman [4] and other authors. In calculations of energy of interaction between bases, especially for configurations identical to those within the nucleic acid double helix, with 3.4 Å between planes of parallel bases, Pullman's group has employed equations of the type (3.4 a-c). Therefore the neglect of repulsion energy E_{rep} is justified. Tables 3.2 and 3.2a give some results for the calculated energies of interaction between base pairs in the double-helix configuration. The energies of stacking interactions, that is of superposed bases are given separately from those of "in (base) plane" interactions of hydrogen bonded bases. U [4] represents the overall interaction energy according to Pullman [4].

TABLE 3.2 **Energies of interaction between base pairs (kcal mol^{-1}); arrows show the $3' \rightarrow 5'$ directions in DNA strand**

Pair	*Stacking interaction*	*Interaction in plane*	U [4]	U' [7]
↑C—G↓ G—C	—11.3	—19.2	—30.5	—15.5
↑G—C↓ C—G	—8.5	—19.2	—27.7	—15.0
↑G—C↓ G—C	—7.7	—19.2	—26.9	—9.5
↑A—T↓ G—C	—9.9	—12.2	—22.1	—13.2
↑A—T↓ C—G	—7.2	—12.2	—19.4	—12.0
↑T—A↓ C—G	—7.0	—12.2	—19.2	—12.1
↑T—A↓ G—C	—7.0	—12.2	—19.2	—13.8
↑A—T↓ A—T	—7.4	—5.5	—12.9	—12.5
↑A—T↓ T—A	—6.1	—5.5	—11.6	—11.4
↑T—A↓ A—T	—5.0	—5.5	—10.5	—11.9

TABLE 3.2a **Energies of interaction between bases in single stranded mononucleotide (kcal mol^{-1}) according to Pullman** [4]

Base	E_{el}	E_{pol}	E_{dis}	U
A—A	+0.67	—0.44	—9.72	—9.49
G—G	+5.12	—1.35	—10.96	—7.19
C—C	+6.69	—1.78	—4.67	+0.24
U—U	+0.49	—0.42	—3.39	—3.32
T—T	+0.55	—0.48	—4.54	—4.51

The last column, U' [7], lists the energies calculated by Poltev and Sukhorukov [7] by neglecting the polarisation term (i. e. $U' = E_{el} + E_{dis} + E_{rep}$); they evaluated $E_{dis} + E_{rep}$ by an equation of type (3.6) with the C-coefficients derived out of a simplified formula (3.4c):

$$C = \frac{3}{2} \cdot \frac{\alpha_\alpha \alpha_{\beta'} I_\alpha I_{\beta'}}{I_\alpha + I_{\beta'}},$$

in which α_x and I_x stand for polarisation and ionisation potential of atom x in the corresponding valence state. The poor agreement of U [4] and U' [7] values for guanine containing base quadruplets, demonstrates the importance of polarisation term, E_{pol}, disregarded by Poltev and Sukhorukov. This finding also agrees with recent calculations [8] which have shown that for distances below 10 Å all terms in equations (3.4 a-c) must be considered for interaction energy (U) calculations.

Bertran's calculations [2] for different reciprocal positions of the two interacting bases have been carried out by equations (3.3 a-c) simplified so as to consider only the interaction of atomic $2p_z$ orbitals of superposed bases (term E_{el} disappears in this approximation); the minimum of the interaction energy U, is obtained for the base pairs at a separation of 3.36 Å, not superposed but deviated from the double helix axis with 45° for E_{pol} and 18° for E_{dis}. This calculation also takes into account the interactions resulting from a charge transfer from occupied molecular orbitals of one base to the empty ones of the other; earlier evaluations of Pullman's [9] have indicated the effect of charge transfer in interactions between bases in DNA double helix to be of only 0.01—0.03 kcal mol^{-1}.

Codon-anticodon interactions between bases of two trinucleotides have been calculated by Claverie [10] according to Pullman's [4] general scheme but with consideration of short range repulsions and computerisation; the energies of interaction between all codons and all anticodons, both correct and false, have been computed. Table 3.3. gives some $\Delta_f U$ differences between the interaction energies of a codon with a false and a correct anticodon. The lowest interaction energies are given by codons with uracyl and adenine and the highest by those with guanine and cytosine. The lowest codon-anticodon interaction energy is —20.3 kcal mol^{-1} for the AUU: UAA pair and the highest, —72.3 kcal mol^{-1} for the GGG:CCC pair. Concerning the anticodons which give small $\Delta_f U$ differences, they seem to predominate for codons rich in A and U, although the AUA and AAA codons, for example, do not have false anticodons with $\Delta_f U$ less than 5 kcal mol^{-1}; however no codon formed only of G and C has false anticodons of $\Delta_f U < 5$ kcal mol^{-1}.

TABLE 3.3 $\Delta_f U$ **differences for interactions of codons with false anticodons and with the correct ones (kcal mol^{-1})**

Codon	*False anticodon*	($\Delta_f U$, kcal mol^{-1})
UUU	AAC (3.2)	CAA (3.8); ACA (4.2)
UUC:	CAG (3.8)	
UGU:	CCA (2.1)	ACC (2.8); CCC (4.9)
AUU:	UAC (3.4)	UCA (5.0)
GGU:	CCC (2.4)	
GUG:	CCC (2.1)	
UGA:	CCU (2.1)	
CGU:	GCC (3.3)	

3.1.2. *Approximations with different potential curves. Polypeptides*

In order to compute the most stable configuration of the polypeptide chain in aqueous solution, one must calculate the interaction energy of different parts of the polypeptide chain with one another and with water molecules of the surrounding solution. Unlike the study of interactions in the nucleic acid double helix, which generally deals with well defined configurations, here one must look for the configuration of lowest free energy from among a huge number of possible configurations. Therefore one must resort to a more empirical approach based upon potential curves for different types of interactions between atoms.

According to Scheraga [11] the computing scheme is as follows: the configurational energy, $W(Q)$ is expressed in terms of generalised coordinates Q which characterise the spatial configuration of polypeptide:

$$W(Q) = U(Q) + V(Q), \tag{3.7}$$

in which $U(Q)$ represents the potential energy of all interpolypeptidic interactions and $V(Q)$ the free energy of all interactions implying the solvent. The role of the solvent, which through the effective dielectric constant D may be sensed in $U(Q)$, and especially in $V(Q)$, will be discussed in some detail in Section 3.2. According to some rules (expressed in Section 5.2), the minima of $U(Q)$ are looked for; the most stable configuration corresponds to the highest partition function, Z:

$$Z = \exp\left(-\frac{U_{min}}{kT}\right)\prod_{i=1}^{3n-6}\left[1 - \exp\left(-\frac{h\nu_i}{kT}\right)\right]^{-1} \tag{3.8}$$

U_{min} is the energy minimum corresponding to the desired configuration, $3n - 6$ the number of vibrational degrees of freedom of polypeptide and ν_i the vibration frequencies of this configuration. There are several rules for the evaluation of the product in equation (3.8). Yet, we recall that the free energy is given by $F = -RT \ln Z$.

$U(Q)$ is calculated as a sum of potential functions of various types of intrapolypeptidic interactions [6]:

$$U(Q) = E_{dis} + E_{rep} + E_{tors} + E_{el} + E_{\mathrm{H}} \tag{3.9}$$

where E_{dis} and E_{rep} represent interactions between non-bonded atoms, usually expressed by potential functions defined like in equation (3.6), E_{el} represents the electrostatic interactions between non-bonded atoms — equation (3.4a) eventually corrected with an effective dielectric constant, E_{tors}, energies associated to rotation around $C_\alpha - C$ and $N - C_\alpha$ bonds expressed, for polypeptide residue i, by:

$$E_{tors}(i) = \frac{1}{2} V_{C_\alpha - C} \cdot (1 - \cos 3\psi_i) + \frac{1}{2} V_{N - C_\alpha} \cdot (1 - \cos 3\varphi_i), \tag{3.9a}$$

in which $V_{C_\alpha - C}$ and $V_{N - C_\alpha}$ are potential barriers to rotations (generally of non-steric nature) and ψ_i and φ_i represent the rotations around C_α—C and N_α—C bonds. $V_{C_\alpha - C}$ and $V_{N - C_\alpha}$ are of the order of a few kcal mol^{-1}. The barrier to rotation about the polypeptidic-partially double C — N bond is of 10 kcal mol^{-1} (as indicated by X-ray studies on lactamic cycles of medium size [12]; for energy calculations the reader is recommended to see the references quoted in [12]). Again referring to equation (3.9), E_H represents the interactions between the hydrogen bonded atoms and may be calculated by an equation given by Stockmayer [13]:

$$E_H = 4\varepsilon\left[\left(\frac{\sigma}{R}\right)^{12} - \left(\frac{\sigma}{R}\right)^{6}\right] -$$

$$-\frac{\mu_{CO} \cdot \mu_{NH}}{R^3}(2\cos\theta_O\cos\theta_H - \sin\theta_O\sin\theta_H\cos\psi). \qquad (3.9b)$$

This equation holds for $\rangle C{=}O \ldots H{-}N\langle$ hydrogen bond, ε and σ are constants corresponding to the last two terms of equation (3.5), μ_{CO} and μ_{NH} are the dipole moments of $\rangle C{=}O$ and $\rangle N{-}H$ bonds, R the separation of O and H atoms, θ_O and θ_H the angles between the two dipole moments and the direction of H bond and ψ the angle between the planes of hydrogen bond $\rangle C{=}O$ dipole and hydrogen bond $\rangle N{-}H$ dipole.

An approach with more detailed potential functions of various type of non-valent interactions is given in Webb's treatise [14]. Table 3.4 lists these functions for the main types of intermoculecular forces (according to Webb's Table 19 [14]). Webb's treatise also lists the experimental data required by calculation of interaction energies (dipole moments, polarisabilities, Van der Waals radii, etc.). The correction factor r in Table 3.4 accounts for the effect of repulsive forces at equilibrium distances which leads to lower absolute values of attraction energies as given by the listed potential functions.

The relative contribution of various types of intermolecular forces to the total interaction energy, was calculated for not too large molecules (see [14], Table 28). For dipolar molecules of moment $\mu < 1$ D, practically the whole interaction energy originates in dispersive forces. Thus for interactions in the gas phase of HCl molecules ($\mu = 1.03$ D) the dipole-dipole interaction contributes with 14.4% to total interaction energy, the dipole-induced dipole interaction with 4.2% and dispersive forces with 81.4%. For NH_3 ($\mu = 1.50$ D) the contributions of the three kinds of interactions are 44.9, 5.3 and 49.8% respectively, for C_2H_5OH ($\mu = 1.66$ D) 55.0, 12.6 and 32.4%, while for H_2O ($\mu = 1.83$ D) 77.4 and 19%. Thus it is only in H_2O that the dipole-dipole interactions bring in nearly 100% of the whole energy. The contribution of dipole-induced dipole interaction is always a small fraction of the overall energy.

TABLE 3.4 **Potential energy functions for different types of interactions**

Interaction	*r*	*Potential function*
ion-ion	0.92	$305 \cdot \frac{Z_1 Z_2}{R_e D}$
ion-dipole	0.83	$57.4 \frac{Z \cdot \mu}{R_e^2 D}$
ion-dipole (random orientation)	0.67	$-1804 \frac{Z^2 \cdot \mu^2}{R_e^4 D^2}$
dipole-dipole	0.75	$-10.8 \frac{\mu_a \cdot \mu_b}{R_e^3 D}$
dipole-dipole (random orientation)	0.50	$-175 \frac{\mu_a^2 \cdot \mu_b^2}{R_e^6 D^2}$
ion-induced dipole	0.67	$-111 \frac{\alpha Z^2}{R_e^4 D^2}$
dipole-induced dipole	0.50	$-7.2 \frac{\alpha \mu^2}{R_e^6 D^2}$
dispersion forces	0.50	$-17.3 \frac{\alpha_1 \alpha_2}{R_e^6} \cdot \frac{I_1 I_2}{I_1 + I_2}$

All equations in Table 3.4 are given for the most favourable spatial orientations. The numerical factors which multiply the potential functions are for conversion into kcal mol^{-1} when Z is the valence of the ions; μ the dipole moments in Debyes (D); R_e the equilibrium distances in Å; α the polarisabilities in Å^3; D effective dielectric constants; I ionisation potentials in kcal mol^{-1}. The potential functions are usually obtained by assuming large R_e distances compared to the diameter of supposedly spherical particles.

3.1.3. *Perturbational approaches and correlations to electronic structure indices*

These have been reviewed for intermolecular forces by Cammarata [15] in connection with theories for the action of drugs. Present theories on drug action claim that the pharmacological response occurs as a result of the combination of drug S with a certain receptor R according to the general scheme:

$$S + R \overset{a}{\rightleftarrows} SR \ldots \overset{b}{\rightarrow} \text{response} \tag{3.10}$$

The response is thus determined directly or mediated by formation of SR complex i.e. by S—R interaction to which contribute modifications of electronic energy (superscript e) of solvation (superscript d), of steric interactions (s) and of conformational energy (p):

$$U_{SR} = \Delta E^e + \Delta E^d + \Delta E^s + \Delta E^p + \text{constant} \tag{3.11}$$

The constant represents other effects. Within a class of drugs it is assumed that the main variations from molecule to molecule are caused by electronic, ΔE^e and solvation ΔE^d terms. According to Klopman and Hudson [16] in the language of molecular orbital theory:

$$\Delta E^e = \sum_{\alpha}\sum_{\beta'}\left[\frac{\gamma_{\alpha'\beta'}}{D}\rho_\alpha\rho_{\beta'} + \sum_{s=1}^{n}\sum_{p'=n'+1}^{\infty}\frac{2c_{s\alpha}^2 c_{p'\beta'}^2}{\varepsilon_s - \varepsilon_{p'}}\beta_{\alpha\beta'}^2\right.$$

$$\left. + \sum_{k'=1}^{n}\sum_{l=n+1}^{\infty}\frac{2c_{k'\beta'}^2 c_{l\alpha}^2}{\varepsilon_{k'} - \varepsilon_l}\beta_{\alpha\beta'}^2\right]. \tag{3.11a}$$

which basically represent (3.3a) and (3.3b) terms transposed in a LCAO theory (linear combination of atomic orbitals) with $\beta_{\alpha\beta'}$ standing for the energy exchange between atomic orbitals on different molecules, ρ_α and ρ_β for net atomic charges, ε_s and $\varepsilon_{k'}$ for the energies of occupied molecular orbitals, $\varepsilon_{p'}$ and ε_l for energies of empty ones, $\gamma_{\alpha\beta'}$, for electronic repulsion integrals (see section 2.6).

Thus, the variation of interaction energy U_{RS} within a class of drugs, S, may be correlated to electronic structure parameters. The equations for ΔE^e and ΔE^d can be expressed in a simpler form if certain assumptions are advanced on the prevailing nature of interaction between R and S (see section 4.4.3).

It is worth mentioning that equations (3.11a) and (3.11b) are also valid if the R—S interaction is chemical, with formation of covalent bonds, in which case ΔE^e and ΔE^d would refer to energy variations connected to the formation of the corresponding activated complex.

3.2. Interactions involving water

Most of the biochemically interesting interactions take place in an aqueous solution and hence calculations of their ΔG must always consider the degree of hydration. Schematically the interaction of molecules $M_1 + M_2$ may be written as:

$$M_1(H_2O)_{\nu_1} + M_2(H_2O)_{\nu_2} \rightleftarrows M_1 \cdot M_2(H_2O)_{\mu} + \frac{\nu_1 + \nu_2 - \mu}{2} H_2O \cdot H_2O, \quad (3.12)$$

if one desires to emphasise special hydrations and interactions of freed H_2O molecules with the remaining solvent. ν_1, ν_2 and μ stand for number of hydration H_2O molecules, for the molecular M_1, M_2 and M_1M_2 species respectively.

The great difficulty with such calculations originates in the structure of liquid water itself. According to Bernal and Fowler [18], water has a fluctuating ice-like and partially destroyed structure. The interaction forces between water molecules — hydrogen bonds — are predominantly Coulombic and determined by the tetrahedral distribution of protons around oxygen. In ice, every H_2O molecule is surrounded by four neighbours; in liquid water the H-bonds are weakened by a slight increase of the average distance between molecules, yet the rotation of some H_2O molecules and their removal from the tetrahedral arrangement causes an increase of the coordination number for some molecules. According to Némethy and Scheraga [19], water contains regions with ice-like structure (about 25—90 molecules between 70°C and 0°C for each region) in thermal equilibrium with "free" molecules of water, of higher coordination number and held together by weaker dipolar forces. One encounters H_2O molecules bonded by 4, 3, 2, 1, or 0 hydrogen-bonds, depending on the position in the cluster (the region with ice-like structure) or in the situation of "free" water. The fraction of molecules unbonded by hydrogen bonds increases from 0.24 to 0.39 between 0°C and 70°C.

This very unstable water-structure is modified in various ways by the substances dissolved therein. Depending on how it interacts with solute particle, hydration water may get more like on ice crystal or more irregular and accordingly the contribution to dissolution entropy is positive or negative. Thus, F^-, OH^-, Ca^{2+}, Mg^{2+} ions increase the degree of ice-like structure — they are "structuring" ions; Cl^-, Br^-, I^-, NO_3^-, SO_4^{2-} and K^+ ions are "destructuring" ones. The ordering effect of ions extends — even though with gradual loss — over several layers of hydration water. The strength of hydration can be measured by means of the rate of exchange of H_2O molecules from within the hydration sphere to within solvent (n.m.r. measurements). Thus, for the exchange of water with "destructuring" ions, the activation energy is negative [20]. The enthalpy effect is always negative, exothermal at hydration of ions. In Table 3.5 the thermodynamic data for the hydration of some ions (according to Webb [14] Table 29) and for some atomic groups met with in biopolymers (according to Gibson and Scheraga [21]) are listed.

If the interaction between two atoms or two groups takes place through one or more layers of water molecules, all electrostatic interactions are modified by the dielectric constant (cf. Table 3.4). Only the short-range dispersion and repulsive forces are not affected, which nevertheless, decrease at such a rate with distance that they are negligible at distances larger than the immediate contact. The difficulty is that the macroscopic value, $D = 80$ (87.9 at 0 °C and 55.9 at 100 °C), is only utilizable for distances larger than 20 Å, whereby even the ion-ion interactions are rarely of any interest [14]. For shorter distances D is smaller. It can be evaluated from the second dissociation constant of dicarboxylic acids — $HOOC(CH_2)_nCOOH$; $n = 0, 1, 2,...$ In most calculations, for very short distances (below 3 Å, yet not at direct contact between ions), one takes $D = 3$, corresponding to ice, wherein the H_2O molecules are fixed, just as in the strong fields of the nearest neighbourhood of ions. For distances between 3 and 7 Å one makes use of equations of the type $D = 6R - 7$ [22]. Other authors, e.g. Scheraga [11], suggest values of D between 1 and 4, for groups separated by an R up to 6 Å, and $D = \infty$ for larger distances. Here also arises the difficulty that the dielectric constant D must be taken smaller in interactions with di- or trivalent ions (stronger field) than in interactions with monovalent ions at the same distance.

Another complication with interactions in solution comes from the ionic atmosphere — the partial arrangement of ions around a given ion, an effect which decreases the potential of the central ion. The potential $V(R)$ at a distance R from the central ion, is:

$$V(R) = \frac{Ze}{RD}\frac{e^{\chi(R_0-R)}}{1+\chi R_0} \therefore \chi = \left(\frac{8\pi Ne^2}{1000DkT}\mu\right)^{\frac{1}{2}}; \quad \mu = \frac{1}{2}\sum_i c_i z_i^2. \tag{3.13}$$

R_0 here is the distance of closest approach of the other ions to the given ion, corresponding to the ionic radius plus the dimension of an H_2O molecule (hydration water) of about 3 Å, or even of two molecules of H_2O for strongly hydrated di- or trivalent ions ($R_0 \cong 10$ Å) [14]. The valence of central ion is denoted by Z, the ions concentrations by c_i, the ionic strength by μ. For water, at 25 °C:

$$\chi = 0.332 \sqrt{\mu} \; (\text{Å}^{-1}). \tag{3.14}$$

Under physiological conditions $\mu = 0.16$, the thickness of ionic atmosphere (which partially screens the potential of central ion, V=0), $1/\chi V$ of 7.5 Å, and the potential energy of atmopshere for a monovalent central ion is of about —0.3 kcal mol^{-1}. This potential energy is lost when an ion is fixed to an enzymic site, the latter uncovering its counter-ionic atmosphere [14]. The direct calculation of energies of interactions with water molecules is difficult. For example, in the calculation of hydration energy of K^+, whose experimental value is (Table 3.5) $\Delta H = -76$ kcal mol^{-1}, Webb [14] takes into account the following interactions: ion-dipole, ion-induced dipole, ion-water and water-water dispersion interaction, and the interaction between the water dipoles,

$$U_{hydr} = -57.4 \frac{nZ_{K^+} \cdot \mu_{H_2O}}{(R + R_1)^2} - 111 \frac{n \cdot \alpha_{H_2O} \cdot Z^2_{K^+}}{R^4} -$$

$$-199 \frac{n\sqrt{8}\alpha_{H_2O}}{R^6}\left(\alpha_{K^+} + \frac{\alpha_{H_2O}}{4}\right) + 14.4 \frac{n\mu^2_{H_2O}}{2\sqrt{2}R^3} =$$

$$= U_{hydr} = -63.1 - 13.8 - 6.03 + 4.24 = -78.69 \text{ kcal mol}^{-1}.$$

Here $n = 6$ (coordination number — first hydration sphere), $Z_{K^+} = 1$, $\mu_{H_2O} = 1.83$ D, $\alpha_{K^+} = 0.87$ Å^3, $\alpha_{H_2O} = 1.44$ Å^3, $R = 2.891$ Å — the sum of the van der Waals radii of K^+ and H_2O-molecule, and $R_1 = 0.274$ Å — the distance between the centre of dipole and the centre of H_2O-molecule. In fact, one should also calculate the interaction with the H_2O-molecules from the next hydration spheres, as well as other effects. A more refined calculation has been performed by Buckingham [23], taking into account the energy of ion-dipole interactions, φ_{dip}, the energy of quadrupole interactions, φ_{quad}, the dispersion interaction, φ_{dis}, the interaction of dipoles and induced multipoles, φ_i, the interaction in the tetrahedral structure of H_2O, φ_l, the interaction with water molecules outside the first hydration sphere, φ_0, the energy required

TABLE 3.5 **Radii, hydration numbers and free energies of hydration for some ions and chemical groups**

Ion, group	—ΔG (kcal mol^{-1})	—ΔH (kcal mol^{-1})	—ΔS (e.u.)	van der Waals radii (Å)	No. of hydration water molecules primarily bonded
H^+	214	223	31.0	—	5
H_3O^+	73	76	10.6	1.64	4
Li^+	111	121	32.1	0.76	6
Na^+	87	95	26.3	1.01	6
K^+	70	76	17.7	1.34	6
NH_4^+	66	70	14.9	1.48	4
Cu^{2+}	473	496	75.0	0.41	12
Zn^{2+}	462	484	74.2	0.57	12
Mg^{2+}	432	456	77.1	0.78	12
Ca^{2+}	352	370	58.4	1.05	10
Ba^{2+}	291	306	48.4	1.39	8
Al^{3+}	888	926	126.8	0.55	24
F^-	113	123	31.8	1.29	5
Cl^-	84	89	18.2	1.81	2
Br^-	77	81	14.4	1.97	2
I^-	89	92	10.1	2.23	1
OH^-	81	91	32.6	1.33	5
H (covalently bonded)	0.62	—	—	1.30	2
O (carbonyl)	3.76	—	—	1.60	4
O (hydroxyl)	5.04	—	—	1.60	6
O^- (carboxyl)	22.40	—	—	1.60	5
N (amide)	1.26	—	—	1.65	2
NH_3^+ (amine)	77.00	—	—	1.75	5
N^+ (imidazole)	9.90	—	—	1.63	3
N^+ (guanidine)	7.20	—	—	1.65	6
S	—1.02	—	—	1.90	6
CH (aliphatic)	—0.26	—	—	1.95	2
CH_2 (aliphatic)	—0.39	—	—	1.95	3
CH_3 (aliphatic)	—1.04	—	—	1.95	8
C (aromatic)	0.22	—	—	1.80	2
CH (aromatic)	0.33	—	—	1.90	3

for creating a "hole" in water structure, φ_h, and the repulsion energy originating in the overlapping of atomic orbitals, φ_r:

$$U_{hydr} = \varphi_{dip} + \varphi_{quad} + \varphi_i + \varphi_{dis} + \varphi_l + \varphi_0 + \varphi_h + \varphi_r.$$

The values calculated for these interaction energies (in kcal mol^{-1}) are: $\varphi_{dip} = -67.7$; $\varphi_{quad} = 26.4$; $\varphi_i = -17.9$; $\varphi_{dis} = -3.7$; $\varphi_l = 1.4$; $\varphi_0 = -38.8$; $\varphi_h = 10.0$; by neglecting the repulsion energy, φ_r, one obtains $U_{hydr} = -90.3$ kcal mol^{-1}, a value slightly more negative than the experimental one.

Under these circumstances, the calculated values of interaction energies with participation of H_2O are rather unreliable. That is why it is preferable to calculate, whenever possible, the interaction energies additively from various types of interaction energies per group; these latter include the effects of solvent energy variations and are available from experimental data.

3.3. Various types of interaction; empirical data

We will observe in the following discussion an empirical classification of the various types of interactions, more detailed than the usual one. We will attempt to group the existing empirical data to make them employable in calculations of overall interaction energies, U, in additive schemes. The empirical data on group-group interaction do implicitly comprise, too, the variations in the solvation energy. More exactly, all that we give below represents free energies of interaction, F or G, at the usual temperature (25 °C, ca. 300 °K).

The interaction types that will be discussed are:

(a) The effect of generating a new particle in solution.

(b) The repulsion forces of the inter-penetration of non-bonded atomic orbitals.

(c) The creation of small "holes" (not filled by other molecules).

(d) Electrostatic interactions between ionised groups.

(e) Hydrogen-bonds (with ionised groups, inclusively).

(f) Dipole-dipole interactions between non-ionised groups and with no hydrogen-bonds.

(g) Hydrophobic interaction (non-aromatic).

(h) Interactions between aromatic groups (contribution to hydrophobic interaction).

(i) Interactions with charge transfer.

All the above-mentioned free energies of interactions refer to direct interactions, at most to interactions between hydrated ions, with only the first hydration sphere rigidly bonded.

As regards the relation between these types of forces and the types E_{el}, E_{pol}, E_{dis} and E_{rep} involved in the perturbation treatment (sections 3.1 and 3.1.1), it is an indirect one. The repulsion forces (b) will be connected to the E_{rep}, yet their effect will only be considered in the determination of closest approach. The ion-ion, hydrogen bonds, electrostatic interactions are the sum of $E_{el} + E_{pol} + E_{dis} + E_{rep}$ wherein, possibly, E_{el} prevails. The contribution of dipole-dipole interactions — as it will be regarded here — would correspond to the contribution of electrostatic term, E_{el}, to the interaction of some un-ionised and not too strongly polar groups. The hydrophobic forces represent a complex effect, a global difference between the interaction of two hydrophobic groups with water and between themselves, plus the interaction of "freed" water with the rest of the solvent. The contribution of the interaction between the aromatic groups and of the charge transfer to the global hydrophobic interaction would correspond to the contribution of conjugated systems of the two partners to the dispersion term, E_{dis}.

3.3.1. *The effect of generating a new parcticle in aqueous solution*

This should be, first of all, a positive entropy of mixing, $\Delta S_{mix} \cong 10$ cal mol^{-1} degree, or a $-T\Delta S_{mix}$ contribution to the free enthalpy G of the interaction of *ca.* — 3 kcal mol^{-1} at 300 °K (*cf.* Appendix 1). Evidently, the same amount of entropy will be lost with the formation of a complex of two particles.

The global change of entropy at dissociation, ΔS, will depend upon other factors, too. First, there appear three new translational and three new rotational degrees of freedom instead of six for vibration and internal rotation. However, since the translation and rotational motions of molecules in solution have also a vibrational character, the contribution to ΔS of this modification of the degree of freedom may be small. Indeed, for the hydrolysis of energetically poor phosphates, $\Delta G \cong -3$ kcal mol^{-1} (section 4.4.2).

The modification of the degree of solvation has an important effect upon dissociation entropy. The formation of some new ionised groups, as a result of dissociation, is expected to cause an important decrease in entropy (*cf.* Table 3.5), which overcompensates for the increase in entropy of mixing. For example, for the dissociation of H_2O into ions,

$$2H_2O \rightleftarrows H_3O^+ + + OH^-,$$

the generation of two new particles in solution corresponds to $\Delta S_{mix} = + 20$ cal mol^{-1} deg^{-1}; the hydration entropies of the ions H_3O^+ and OH^- —10.6 and —32.6 cal mol^{-1} deg^{-1} respectively; the sum of these effects is of —23.2 cal mol^{-1} deg^{-1}, close enough to the experimental entropy, —18.8 cal mol^{-1} deg^{-1}, of this dissociation process.

3.3.2. *The short-range repulsion forces*

These are usually considered equal to zero in empirical determinations of interaction energies, even though they actually diminish a little the attractive interactions. These forces are extremely important in the determination of contact distances between various groups. In a first approximation one may consider that the distance between atoms in direct contact is equal to the sum of van der Waals radii of these atoms. This principle lies at the basis of molecular models of stereochemistry. In Table 3.6 are given the van der Waals radii and radii for covalent bonds of various atoms and groups of atoms (most of them according to [14], Tables 20 and 33). For other groups van der Waals radii can be calculated by means of bond lengths, valence angles and van der Waals radii of peripheral atoms. For a lot of atoms, the van der Waals radii are given in Table 3.5. The precision to which one can calculate the distances between the nuclei of atoms in van der Waals contact is of some tenths of Å.

The repulsion energy at steric non fit depends, evidently, on the rigidity of the molecules involved. Thus, the affinity of benzoic acid, $C_6H_5COO^-$, for the antibody elicited to p-(azophenylazo)benzoate is decreased, if one introduces ortho-substituents into benzoic acid, with the following values (in kcal mol^{-1}): CH_3 1.94; Cl 1.22; I 2.40; NO_2 and COO^- greater than 2.5. The $C_6H_5CH_2COO^-$ anion does not bind at all to this antibody ([14], [58]).

TABLE 3.6 **Covalent and van der Waals radii**

Atom (group)	*Covalent radius* (Å)		*van der Waals radius* (Å)
	single bond	*double bond*	
H	0.30	—	1.2
C	0.77	0.67	1.57
N	0.70	0.61	1.5
O	0.66	0.57	1.4
S	1.04	0.95	1.85
P	1.10	1.00	1.9
As	1.21	1.11	2.0
F	0.64	0.55	1.35
Cl	0.99	0.90	1.80
Br	1.14	1.05	1.95
I	1.33	1.25	2.15
CH_3	—	—	2.0
C_6H_5 (aromatic rings half-thickness)			1.7—1.85
—OH	—	—	1.60
$—NH_2$	—	—	1.85
$—NO_2$	—	—	2.02
$—COO^-$	—	—	2.10
$—NMe_3^+$	—	—	3.17
$—AsMe_3^+$	—	—	3.68
$—NEt_3^+$	—	—	4.17

3.3.3. *The creation of a small hole*

In interactions between biomolecules we come across situations when cavities are formed which are too small to accommodate water molecules (van der Waals radius of ca. 1.5 Å) or small groups, like CH_3 (van der Waals radius cf ca. 2.0 Å) but are large enough to cause an important weakening of intermolecular forces. The equilibrium intermolecular distances between nuclei of the atoms in contact are of 3—4 Å; the increasing of this distance from 3.5 Å to 5 Å decreases the dipole-dipole interactions (proportional to R^{-3}) by a factor of *ca.* 3, and the dispersion interaction energy (proportional to R^{-6}) by a factor of *ca.* 9. One may assume, in a fairly good approximation, that such cavities cause the interaction energy between atoms on the opposite sides of the cavity to vanish.

In this case, the situation would be analogous to the creation of a separating medium-vacuum interface. The free enthalpy for the formation of this cavity, ΔG_{cav} can be evaluated by means of the surface energy required for the creation of σ-surface:

$$\Delta G_{cav} = N_A \gamma \sigma \quad \text{or} \quad \Delta G_{cav} = N_A \left(\frac{\gamma_{HF}}{2} + \frac{\gamma_{Pol}}{2} - \gamma_{HF,Pol} \right) \sigma, \qquad (3.15)$$

as either both faces of the cavity are hydrophilic or hydrophobic or one is hydrophobic and the other hydrophilic (polar); in the latter case $\gamma_{HF,Pol}$ represents the surface tension at the hydrophobic-polar media interface.

The surface tension γ of hydrophobic substances is of 20—30 dynes cm^{-1}, at usual temperatures (cyclohexane, 20.5; benzene, 28.9; methanol, 23 dynes cm^{-1} at 20 °C). For polar substances one may take for example water, with $\gamma \cong 70$ dynes cm^{-1}; alcohols exhibit towards outside a hydrophobic methyl group. Therefore, for the creation of a small cavity with a surface area of the opposite sides $\sigma = 10$ Å^2, in a hydrophobic medium with $\gamma \cong 25$ erg cm^{-2}, a surface work of $\gamma\sigma = 2.5 \times 10^{-14}$ erg is required, or for a mole of such cavities, $\Delta G_{cav} = 0.35$ kcal mol^{-1}. In a polar medium, $\gamma = 70$ erg cm^{-2}, $\Delta G = 1$ kcal mol^{-1} for such cavities ($\sigma = 10$ Å^2 = 2 faces of 2.2 Å × 2.2 Å each).

3.3.4. *Electrostatic interactions between ionised groups*

Most biological processes take place at pH values of *ca.* 7. In Table 3.7 are given p*K*-values for the nucleotides and native amino acids (according to [24], [25]), as well as for some other substances of biochemical interest. For pH > p*K*, the groups are in deprotonated form, while for pH < p*K* they are in protonated form. Thus at pH = 7 in nucleic acids only the phosphate groups are ionised (anions); in amino acids the —COOH and —NH_2 groups are found as anions and cations, respectively, while for the amino acid side chains, considering their isoelectric pH, the carboxyl groups of aspartic and glutamic acids are founds as anions, the amino groups of lysine and arginine as cations. Histidine is predominantly cationic too (*ca.* 80%).

As regards the free enthalpy of the anion-cation bond in aqueous solution, it is much smaller than in crystals: salts in aqueous solutions (at concentrations of *ca.* 1 M) are almost entirely dissociated. The experimental values for this ΔG, related to anionic and cationic groups of biological importance have, in general, been obtained by studying the competitive equilibria for a given site (e.g. an antibody) between two

TABLE 3.7. **pK_a values for biochemically important substances**

Substance		pK_a (group)	
Adenine		4.15; 9.80	
Guanine		3.3; 9.2	
Cytosine		4.60; 12.16	
Uracil		9.45; $>$ 13	
Thymine		9.94; $>$ 13	
Adenylic acid	0.9 (RPO_4H_2);	3.7 (—NH_2);	6.0 (RPO_4H^-)
Guanylic acid	0.7 (RPO_4H_2);	2.4 (—NH_2);	6.0 (RPO_4H^-); 9.3 (OH base)
Cytidylic acid	0.8 (RPO_4H_2);	4.2 (—NH_2);	6.0 (RPO_4H^-);
Uridylic acid	1.0 (RPO_4H_2);		5.9 (RPO_4H^-); 9.3 (OH-base)
Guanine (in DNA)	2.35—2.9 (—NH_2);		
Adenine (in DNA)	3.7—3.85 (—NH_2);		
Cytosine (in DNA)	4.5—4.85 (—NH_2);		
Thymine (in DNA)		11.4 (—CONH—)	
Amino acids	1.77—2.58 (α—COOH);	8.5—10.60(α — NH_2);	pI = 5.74—6.3
Amino acid side chains			
Cystine	1.02 (—SH → —SH_2^+);	8.00 (—SH → S^-);	pI = 5.02
Tyrosine	10.87 (phenolic OH)		pI = 5.63
Histidine	6.10 (imidazolic = N—)		pI = 7.64
Lysine	10.53 (NH_2)		pI = 9.47
Arginine	12.78 (guanidinic NH_2)		pI = 10.76
Aspartic acid	3.86 (ω—COOH)		pI = 2.98
Glutamic acid	4.07 (ω—COOH)		pI = 3.08
CH_3COOH	4.76 (COOH)		
$CH_3CHOHCOOH$	3.78 (COOH)		
H_3PO_4	2.14; 7.13; 12.39		
$H_4P_2O_7$	0.85; 1.98; 8.44		

molecules of similar shape and size, of which only one bears an ionised group (e.g. antigenic determinants one carrying a CMe_3 group while the other a NMe_3^+ or NO_2 and COO^- groups). From this kind of study a $\Delta G = -1.2$ kcal mol^{-1} was calculated for the anion-cation bond in NMe_3^+ and a value of —4.8 kcal mol^{-1} for COO^- [27]. From studies of bonding of some inhibitors or ionic and non-ionic (otherwise similar) substrates to acetylcholinesterase, which probably has a COO^- group in its active centre, one obtains for the bond between carboxylate and ions of quaternary ammonium, values between 1.2 and 2.1 kcal mol^{-1} for ΔG [28]. Thus, for the interaction between organic ammonium and carboxylate ions, the average value is *ca.* $\Delta G = -1.7$ kcal mol^{-1}. The relative strength of the ionic bonds appears to decrease in the order: guanidinium (arginine) $>$ ε — amino(lysine) $>$ end group ammonium $>$imidazolic N (histidine) [29].

The specific bonding of some alkali or alkaline earth ions —e.g. in kryptates synthesised by Lehn [30] (some diazapolyoxanic macrocycles with molecular weights of *ca.* 500—1000 daltons) —is probably due to the radius of cation fitting that of the relatively rigid anionic cavity, so that neither a deformation of the anionic cavity, nor of the remaining empty cavities is required. As regards the cation, the fit may also refer to the cationic radius plus that of first hydration sphere [31]. This kind of consideration also explains the selective permeability of biological membranes to certain ions. Metallic ions, especially those of the transitional series, can become specifically bonded to certain groups, by means of complexing effects. By this complexing effect, complexes—sometimes fairly stable—in which two biomolecules are involved, may also be formed. By this mechanism some dyes are attached to natural or synthetic fibres (for example, with Al^{3+} as mordant).

The stability of such complexes is, primarily, dictated by the number of coordination points within the complexing site of the biomolecule, then by the strength of this site and by how well the radii of site and ion fit. A large number of coordination points per site means a small loss of entropy of mixing on complex formation. Thus, the haem or the porphyrinic ring in chlorophyll, having four coordination sites, strongly bonds to Fe^{2+} and Mg^{2+} ions, respectively. In order that two biomolecules be complexed through an ion, each biomolecule must appear with a site of 2—3 points of coordination. The cation-ligand interaction is governed by the principle that cations of the "hard" Lewis acid type prefer ligands of the "hard" Lewis-base type, while "mild" acids prefer "mild" bases (cf. e.g. [55]). According to increasing "hardness", the usual cations may be classified this way (in H_2O): ("mild") Hg^{2+}, Ag^+, Cu^+, Hg^+, Cu^{2+}, Na^+, Fe^{2+}, Fe^{3+}, Co^{2+}, Ni^{2+}, Ca^{2+}, Mg^{2+}, Al^{3+} ("hard"). As to the usual ligands, the order of increasing "hardness" (in water) is: ("mild") I^-, HS^-, CN^-, Br^-, Cl^-, pyridine, aliphatic amines, H_2O, HO^-, COO^-, PO_4^{3-}, F^- ("hard"). Thus, mercury does preferentially bind SH-groups, while cations such as Ca^{2+}, Mg^{2+}, Al^{3+} prefer carboxylic or phosphate groups.

Another effect occurring on formation of complex compounds, is given by the cleavage parameter Δ depending on both ligand and central cation. The effect only appears for cations with incomplete *d* shells and is particulary strong for cations with three or six *d* electrons, all of which can be placed on the three bonding *d* orbitals (cf. section 2.1) [55]. The ligand electron donating atoms, are—according to increasing Δ—as follows:

$$I < Br < Cl < S < F < O < N < C,$$

while the transitional cations (numerals in parantheses represent the number of *d* electrons):

$$Mn^{2+}\ (5) < Ni^{2+}\ (8) < Fe^{2+}\ (6) < Fe^{3+}\ (5) < Cr^{3+}\ (3)$$
$$< Cr^{3+}\ (3) < Co^{3+}\ (6) < Pd^{4+}\ (6) < Pt^{4+}\ (6).$$

Thus, the Co^{3+} ion prefers to bond to aliphatic amino-groups, but since these are protonated at pH = 7 — it bonds to the free electrons of the N-atom (pyridinic) in the purinic and pyrimidinic heterocyclic molecules.

As to preference for ligand atoms, Na^+, K^+, Mg^{2+}, Ca^{2+}, Mn^{2+} and Mn^{3+} usually prefer —O^-; Fe^{2+}, Fe^{3+}, Co^{2+} and Co^{3+}, prefer —O^- or $\gtrdot$N; Cu^+, Cu^{2+}, Zn^{2+} prefer $\gtrdot$N or — S^- while Mo^{5+} and Mo^{6+} prefer —S^- [59]. The bonding of cations (inorganic and organic) to native DNA presents some base specificity: dG.C rich DNA's bond Ag^+ stronger than dA.T rich DNA' s, Hg^{2+}, Me_4N^+, and $EtNMe_3^+$ prefer dA.T rich DNA's while Na^+, Li^+, Cs^+, K^+, Ca^{2+}, arginine, lysine, spermine bond equally strongly to DNA of any base composition. The preference, wherever it exists, is weak—a ΔG of *ca.* 0.2 kcal mol^{-1} is released on transfering a Me_4N^+ ion from a dG.C to a dA. T sequence. Such a preference has nothing to do with hydrophobicity but rather with the higher deformability of dA.T rich DNA's: Me_4N^+ shows the same preferential bonding to DNA rich in dA. U base pairs [70].

One should mention that hydrogen bonding may strengthen electrostatic bonding getting to the so called salt bridge: that one between the carboxylate ion of Asp-194 and the α-ammonium ion of Ile-16 in chymotrypsin has a $\Delta G = 2.9$ kcal mol^{-1}, a $\Delta H = -7.1$ kcal mol^{-1} and $\Delta S =$ $= -27$ e.u. [62] (see section 3.3.5).

Finally it is worth mentioning that special problems may arise in polyelectrolytes; these are reviewed by Brenner and McQuarril [60] and Katchalsky [61].

3.3.5. *Hydrogen bonds*

Hydrogen bonds are formed between an X—H bond sufficiently strongly polarised (X electronegative enough) and the pair of free electrons (not involved in conjugation) of an atom Y suitably electronegative:

$$X-H\cdots :Y$$

Table 3.8 lists the average bonding energies ΔH for various kinds of hydrogen bonds (according to Fergusson [32] p. 128). The enthalpy effect ΔH for hydrogen bonds is negative but the entropy effect ΔS is also negative so that, over all, ΔG is close to zero for hydrogen bond formation. In Table 3.9, data for several substances are given, according to Pimentel and McClellan [33] (Table 3.7—X and Appendix B) and according to [63]; the errors are of *ca.* ± 1 kcal mol^{-1}.

The possibility of H-bond forming means a relatively good solubility in water. According to this criterion, hydrogen bonds do form between O—H···O, N—H···O, F—H···O, O—H···N pairs but not with the halogens covalently bonded to carbon (fluorine carbides are insoluble in water), neither with atoms Y from the third and higher periods (e.g. with S), except for relatively weak hydrogen bonds. C—H··· Y bonds may be formed whenever on atom C there are strongly electronegative substituents; these bonds seem weak, e.g. $CHCl_3$ is poorly water-soluble. For atom Y the directing of free atomic orbital seems to matter too: according to a decrease of degree of direction, that is of hydrogen bonding tendency, the hybrid orbitals stay in the sequence: $sp^3 > sp^2 > sp > p > s$. Hydrogen bonding is favoured by conjugation effects which entail formation of

TABLE 3.8 **Average bonding energies for various types of hydrogen bonds.**

X—H Y	$-\Delta H$(kcal mol^{-1})	*Sample system*
F—H··· F	7	
N—H··· F	5	
O—H··· N	7	
O—H··· O	6	
N—H··· O	2.3	aniline-benzophenone
N—H··· N	2—4	ammonia crystals
C—H··· O	2.6	$(CH_3)_2CO.HCCl_3$
O—H ··· π-system	2—4	Alcohol-double bond

TABLE 3.9 **Thermodynamic data for hydrogen bonds (data per bond ; ΔH and ΔG (300 K) in kcal mol^{-1} ; ΔS in cal mol^{-1} deg^{-1}**

X—H Y	*System*	$-\Delta H$	$-\Delta S$	ΔG
O—H ··· O	CH_3COOH	7.0	18	—1.6
	H_2O	5.0	25.8	+2.7
	CH_3OH	4.5	20	+1.5
	C_6H_5OH	4.3	25	+2.2
N—H ··· O	$CH_3CONHCH_3$	3.9	13	0
	diphenylamine-dioxane	2.3	16	+2.5
	diphenylamine-$HCON(CH_3)_2$	3.6	16	—1.2
	δ-valerolactam	5.15	11	—1.8
N—H ··· N	NH_3	4.4	26.8	+3.6
	CH_3NH_2	3.4	22.0	+3.2
F—H ··· F	HF	6.8	3	—5.9
C—H ··· N	HCN	4.0	16.5	+1
S—H ··· O	$C_6H_5SH + (CH_3)_2CO$	3.2	13	+0.7
S—H ··· N	C_6H_5SH + pyridine	3.4	11	+0.9
O—H ··· S	$C_6H_5OH + (C_2H_5)_2S$	3.4	8	—1.0

partially positive charges on atom X and partially negative on Y (e.g. urea).

The data in Table 3.9 refer to hydrogen bonds in non-polar solvents. In aqueous media, the hydrogen-bonding also implies a rearrangement of hydrogen bonds in H_2O itself:

$$\begin{aligned} &X—H \cdots OH_2 + —Y: \cdots H_2O \rightarrow \\ &\rightarrow X—H \cdots :Y— + H_2O \cdots H_2O \end{aligned} \tag{3.16}$$

For example, the dimerisation of $CH_3CONHCH_3$ in aqueous solution (one H-bond dimer^{-1} is formed) has at 25 °C, $\Delta H = 0$ kcal mol^{-1}, $\Delta S = -10$ cal mol^{-1} deg^{-1}, $\Delta G = +3.1$ kcal mol^{-1} [34]. The forming of hydrogen bonds with H_2O is favoured in addition to the ΔG values from Table 3.9, by the high concentration of liquid H_2O (of *ca.* 55 mole l^{-1}), thus by a term - RT ln $c_{H_2O} \cong -2.4$ kcal mol^{-1}. Hence, according to Scheraga [11], for the interaction of H_2O with the —CONH group in polypeptides, the free energy of the hydrogen bond can be calculated according to:

$$\Delta G = -2RT \ln (1 + Ka_{H_2O}) .\because$$

$$Ka_{H_2O} = \exp \left(- \frac{\Delta H' - T\Delta S'}{RT} \right), \tag{3.17}$$

and from the levelling of melting curves of polyalanine in H_2O, $\Delta H' =$ -2.74 kcal mol^{-1}, $\Delta S' = -5.78$ cal mol^{-1} deg^{-1}. At 300 K one obtains, thus, $\Delta G = -1.6$ kcal mol^{-1}. For the OH (tyrosine)-carboxylate bond, Scheraga [36] found $\Delta H = -7.7$ kcal mol^{-1}, $\Delta S = -24$ e.u., $\Delta G = -0.55$ kcal mol^{-1}, and for hydrogen bonding between urea molecules in water $\Delta H = 1.4$ kcal mol^{-1}, $\Delta G = -0.4$ kcal mol^{-1} [39]. A large number of base pairing between purine and pyrimidine bases in aqueous solution has been studied by Nakano and Igarashi [40]; the association constants range between 3 M^{-1} and 80 M^{-1}, corresponding to $-\Delta G$ between 0.7 and 2.6 kcal mol^{-1}. For example, the association constants for the AU and CG-pairs in H_2O are of 4.16 M^{-1} and 3.32 M^{-1} respectively; $K = 80$ M^{-1} between adenine and 8-methylcaffeine. In water, the SH and SCH_3 groups seem to yield, by hydrogen bonding, $-\Delta G$ of *ca.* 0.7 and 1 kcal mol^{-1} compensating the hydrophobic effect (section 3.3.7) respectively.

The hydrogen bond produces a decrease in ΔS, owing, first of all, to the entropy of mixing; the forming of a second, of a third hydrogen bond between a pair of associated molecules entails a slow down in the decrease of ΔS. At limit, for the forming of a hydrogen bond, between a pair of already fixed partners, $\Delta H \cong 0$, the energy of the newly formed hydrogen bond being approximately compensated by the breakdown of the hydrogen bonding with H_2O, while ΔS corresponds to the increase of entropy for freeing of two water molecules. One may, thus, expect a fairly negative ΔG with the interaction of e.g. two amide groups; this may also be seen from the relatively negative ΔG for the interaction between the molecules of acetic acids (Table 3.9); the first of the hydrogen bonds might have the characteristics of methylacetamide, while the second would correspond to ΔG of -3 kcal mol^{-1}. Also, for the formation of the complex AU from A and U in $CHCl_3$, $\Delta G = -2.7$ kcal mol^{-1} [38]; between these bases two hydrogen bonds are also formed. Another case is that of the hydrogen bonding between the two strands of a nucleic acid double helix. The hydrogen bonding groups are directed towards the interior of the already dehydrated helix and the "crystalline" structure reduces the freedom of motion, so that $\Delta S \cong 0$, $\Delta G' \cong \Delta H$, for the formation of a hydrogen bond. In case two X—H···H—X or Y···Y groups are side by side, even with disregarding of steric repulsion, one gets $\Delta G'' \cong$ $\cong -2\Delta H$ that is of the order $+10 \ldots +15$ kcal mol^{-1}. This explains how hydrogen bonding, although contributing with $\Delta G \cong 0$ to the formation of

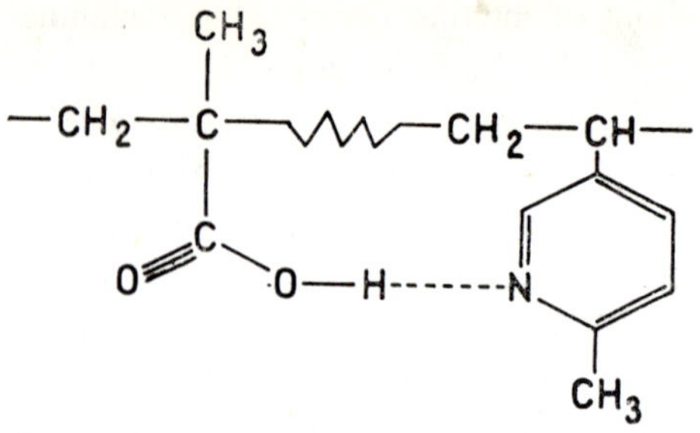

Figure **3.1.** Internal hydrogen bond.

a double helix in the nucleic acid random-coil, may have an important share in the observing of complementarity; a false base-pairing will raise total ΔG of the double helix by 1—2 $\Delta G''$ units.

Hydrogen-bonds can strenghten sensibly the interaction between ionised groups of opposite charges. Thus interactions between carboxyl and ammonium groups (not quaternary ones) imply also the formation of N—H···O hydrogen bonds. For example, the formation of a saline bond of the copolymer [35] in Figure 3.1 in aqueous solution requires $\Delta G \cong -6$ kcal mol^{-1}. The data in Table 3.9 referring to interactions in nonpolar solvents, do automatically include the difference between the interaction of X—H···Y groups in the actual hydrogen bond and the interactions of these groups, separately, with the hydrophobic medium. Therefore, according to the arguments above, the difference, ΔG, between the free enthalpy of a X—H or Y group in a cavity with a (Y or X—H respectively) partner and that of the same group in a hydrophobic cavity is of the order of ΔH for X—H···Y interaction, as measured in a hydrophobic medium; the freedom of motion is reduced to approximately the same extent in both cavities, so that the entropy difference is $\Delta S' \cong 0$.

3.3.6. *Uneven charge distribution, dipole moments*

These must not be implied in hydrogen bond formation, and may bring about an appreciable negative contribution to the free energy of interaction, ΔG, especially in the case of superposed planar aromatic cycles [37]. The contribution U_{dip}, of dipole interactions between two aromatic superposed cycles of dipole moments μ_1 and μ_2 Debyes, forming an angle θ, at equilibrium separation of $R = 3.5$ Å is, in kcal mol^{-1}, given by equation (3.18a)

$$U_{dip} = \frac{\mu_1 \mu_2 \cos\theta}{R^3} \quad (3.18) \qquad U_{dip} \cong 0.34\,\mu_1\mu_2 \cos\theta \left(\frac{\text{kcal}}{\text{mol}}\right) \quad (3.18a)$$

In Figure 3.2 are shown the approximate directions of dipole moments in aromatic side chains of amino acids and nucleic acid bases, according to charge distributions in Figure 2.8. (and Figure 2.3.) together with the numerical values of dipole moments. A more exact evaluation of the interactions between electrostatic charges is found in sections 3.1.1. and 3.1.2.

Experimental determinations of such dipole interactions do not exist as the association constants measure the overall effect of so called stacking interactions, van der Waals interactions and dipole-induced dipole interactions. Dipole-induced dipole interactions may become quite strong.

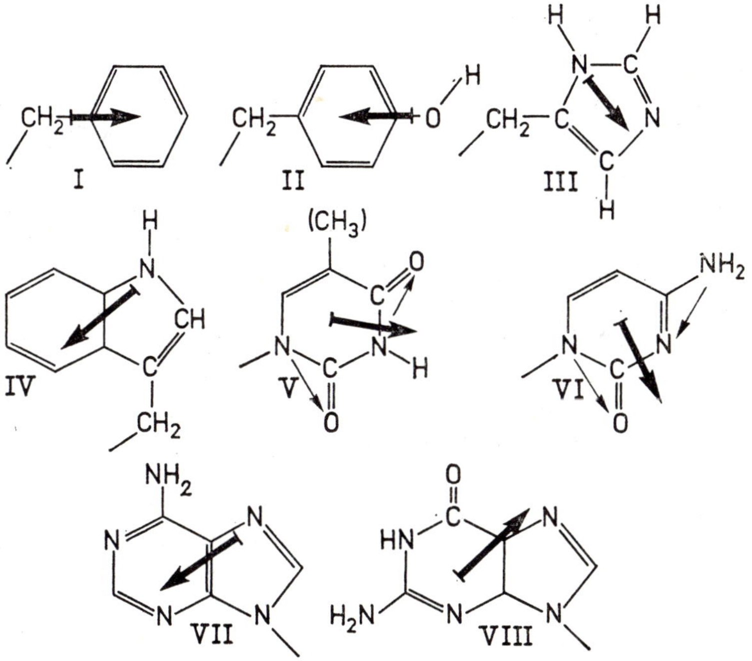

Figure 3.2. Approximate directions of dipole moments for aromatic cycles in amino-acidic side chains and nuclei bases (+ → —)
I: Phenylalanine ($\mu = 0.4$ D); II: thyrosine ($\mu = 1.5$ D); III: histidine ($\mu = 4.0$ D); IV: triptophan ($\mu = 2.0$ D); V: uracyl(thymine) ($\mu = 3.9$ D); VI: cytosine ($\mu = 8.0$ D); VII: adenine ($\mu = 2.8$ D); VIII: guanine ($\mu = 6.8$ D).

The benzene-water and hexane-water partition constants P, of ethyl-alcohol [43] ($\log P = -1.5$ and -2.25 respectively; $\mu_{C_2H_5OH} = 1.7$ D) may be considered to differ because of dipole (alcohol)-induced dipole (benzene; polarisability $\alpha = 24$ ml mol^{-1}) interactions. The contribution of these interactions, as it comes out of log P difference, should amount to *ca.* 1.0 kcal mol^{-1}. According to Table 3.4 these interactions are proportional to the polarisability and the square of the dipole moment.

3.3.7. *Hydrophobic forces*

Hydrophobic forces represent a complex effect whose main mechanism seems to be the reduction of number of water molecules in direct contact with hydrophobic surfaces; the possibilities of hydrogen bond formation are so reduced:

$$A_n \cdot H_2O + B_m \cdot H_2O \rightarrow A \cdot B + \frac{n+m}{2} H_2O \ldots H_2O \qquad (3.19)$$

Only weaker dispersion forces are exerted between the other parameters (A with B and A, B with H_2O). The result of these complex interaction is a structural rigidity of water molecules near the hydrophobic surface which makes them more like ice crystals. Therefore, the entropy of formation of hydrophobic bonds between two groups A and B (of the —CH_2-type, etc.), i.e. for reaction (3.19), is positive, $\Delta S = +1.7 \ldots + 11$ e.u., $\Delta H = + 0.3 \ldots 1.8$ kcal mol^{-1}, $\Delta G = -0.2 \ldots -1.5$ kcal mol^{-1} [41]. Interactions of the same type, solvophobic interactions, occur in other solvents such as glycerine, formamide, ethylendiamide, propane-1,3-diol, etc., whose molecules have two or more hydrogen-bond forming centres (both donating and accepting) and so are able to form three-dimensional networks through hydrogen bonds [42]. The free enthalpy for hydrophobic interaction may be deduced from a study of equilibrium of partition of molecules between water and a hydrophobic solvent, most often octanol (naturally saturated with water) [43]. The partition coefficient P_i, of a substance between water and the solvent "i" may be defined as:

$$P_i = \frac{\text{Equilibrium concentration of substance in solvent } i}{\text{Equilibrium concentration of substance in water}} \qquad (3.20)$$

$$\Delta G_i = RT \ln P_i = -1.4 \log P_i \text{ kcal mol}^{-1}, \text{ at } ca.\ 300°\text{K}$$

The partition coefficient P_i and the free enthalpy change ΔG_i corresponding to a substance passing from water into the hydrophobic solvent do not depend strongly on the latter (ethanol or other alcohols, acetone, dioxane, etc.) [44]. Thus, the passage of a CH_4 molecule from benzene, diaethyl ether or carbon tetrachloride into water, entails a change in ΔG of +2.6, +3.3 and +2.9 kcal mol^{-1} respectively, while for the passage of ethane from benzene or carbon tetrachloride into water, the change is of the value of +3.8 and +3.5 respectively (all values at 25 °C). For the partition coefficients of a substance (practically of any substance), "K" between water and two hydrophobic solvents ("1" and "2") linear relationships of the type [43]:

$$\log P_2(K) = a_{12} \log P_1(K) + b_{12} \tag{3.21}$$

may be found, in which a_{12} and b_{12} are coefficients characteristic to the pair of solvents. Such solvent regression equations are usually rather precise, but hydrogen-donor and hydrogen-acceptor must be, for most solvent pairs, correlated by two different equations. As an example, for the two types of solutes, the regression equations between partition coefficients in benzene $P_{C_6H_6}$, and octanol, P_{oct} are [43] (with r standing for linear correlation coefficients and N for the number of solutes tested):

hydrogen-donor solutes:

$$\log P_{C_6H_6} = 1.015 \log P_{oct} - 1.402 \quad r = 0.962 \quad N = 33 \tag{3.21a}$$

hydrogen-acceptor solutes:

$$\log P_{C_6H_6} = 1.223 \log P_{oct} - 0.573 \quad r = 0.958 \quad N = 19 \tag{3.21b}$$

For the significance of coefficient r, see Appendix 2. For hydrophobic interactions, ΔG is additive over atomic groups and substituents within a molecule, so that Hammett type equations are valid. Characteristic "π" hydrophobicity constants were introduced by Hansch *et al.* [43] for various groups. The π_X constant of substituent X may be determined from partition coefficients of RX and RH molecules:

$$\pi_X = \log P(RX) - \log P(RH) \tag{3.22}$$

π_X values are strongly affected by conjugation effects, by the aliphatic or aromatic nature of the R residue and by the folding of aliphatic chains. The π_X^{ar} values for an aromatic residue R are usually higher than π_X^{al} for aliphatic residues; some $\pi_X^{ar} - \pi_X^{al}$ differences are listed in Table 3.10; the branching and folding of aliphatic chains influences these values. Table 3.10 lists all these effects and may be used in calcu-

TABLE 3.10 **Pertinent parameters for hydrophobic interaction of different X groups**

Structural element	$\pi_X \pm \Delta\pi_X$	$\pi_X^{ar} - \pi_X^{al}$	G_X kcal mol^{-1}	$\Delta(\Delta G)_X$	R_X cc mol^{-1}	$\frac{-G_X}{-R_X}$
$-CH_2-$	0.50 ± 0.02		—0.70	—0.49 *)	5.08	0.138
Ramification:						
a) of $(CH_2)_n-$ chain	—0.20 ± 0.02		+0.28			
b) of functional group	—0.20 ± 0.05		+0.28			
c) ring closure	—0.09 ± 0.02		+0.13			
Double bond	—0.30 ± 0.05		+0.42			
Folding of $(CH_2)_n-$ chain	—0.60 ± 0.05		+0.84			
Intramolecular H-bond	—0.65 ± 0.10		—0.91			
—COOH	—0.65 ± 0.03	0.39	—0.19	(—0.52**)	6.05	
—OH	—1.16 ± 0.03	0.49	+1.62	(—0.57)	3.26	
$-NH_2$	—1.16 ± 0.03	—0.04	+1.62	(—0.29)	5.15	
>C=O	—1.21 ± 0.03	0.16	+1.70		4.50	
—CN	—0.84 ± 0.04	0.26	+1.18		5.89	
—O—	—0.98 ± 0.05	0.45	+1.38		1.48	
$-CONH_2$	—1.71 ± 0.05	0.22	+2.40		9.65	
—F	—0.17 ± 0.03	0.31	+0.24		1.06	
—Cl	0.39 ± 0.04	0.32	—0.55	—0.85	5.38	0.086
—Br	0.60 ± 0.04	0.26	—0.84	—0.89	9.19	0.092
—I	1.00 ± 0.05	0.12	—1.40	—1.10	14.08	0.100
$-NO_2$	—0.85 ± 0.05	0.57	+1.20	(—0.79)	6.42	—

The basic residue R is aliphatic.
G in kcal mol^{-1} . *R* values in cm^3 mol^{-1}.

*) refers to CH_3.
**) refers to $-COO^-$. Values in parantheses are not purely hydrophobic; hydrogen bonds also interfere.

lations, by additivity rules, of hydrophobicity constants and partition coefficients of large and complicated molecules, according to Hansch and Elkins [43]. Also given are the molar group refractivities R_X, and the contribution, $\Delta(\Delta G_X)$, of the respective group to the interaction of some antibodies elicited by *p*-azophenylarsonate and *p*-phenylazobenzoat coupled to substitution derivatives of these molecules (according to [14], Tables 17, 33). The values of $G_X = 1.4\ \pi_X$ (corresponding to the free enthalpy of hydrophobic interaction) and G_X/R_X ratios are also listed.

Table 3.11 gives the G_X contributions of amino acid side chains to hydrophobic interactions according to Tanford [44] (some values were found in references [45] and [46]), together with side chain volumes V_X (in Å^3) [47] and molar refractivities (cm^3 mol^{-1}). A certain parallelism is evident between G_X and V_X or G_X and R_X of hydrophobic groups, i.e. without hydrogen bonding ability. The G_X/R_X ratio is approximately

TABLE 3.11 **Hydrophobic interaction parameters for amino acid side chains**

Amino acid	*Side chain*	$-G_X$(kcal mol^{-1})	V_X(Å^3)	R_X(cm^3 mol^{-1})	$-G_X/V_X$	$-G_X/R_X$
Gly	H	0	36.3	1.07	0	0
Ala	CH_3	0.73	52.6	6.15	0.0139	0.119
Val	$(CH_3)_2CH$, C_3H_7	1.69	85.1	15.73	0.0198	0.107
Leu	$(CH_3)_2CHCH_2$, C_4H_9	2.49	102.0	20.81	0.0238	0.116
Ileu	$CH_2CH_2CH(CH_3)_2$, C_4H_9	2.97	102.0	20.81	0.0290	0.142
Phe	$CH_2C_6H_5$	2.65	113.9	20.81	0.0232	0.086
Pro	pyrolidine C_4H_8N	2.60	73.6	~25	0.0353	0.104
Met	$CH_3SCH_2CH_2$, C_3H_7S	1.30	97.7	24.29	0.0133	0.054
Cys [46], [47]	$HSCH_2$	0.70				
		1.00	68.3	14.13	0.0103	0.050
Trp	Indole C_9H_8N	3.00	135.4	~46	0.0221	0.065
Tyr	$CH_2C_6H_4OH$	2.97	116.2	~33	0.0256	0.090
His	Imidazole	1.10	91.9			
Threo	CH_3CHOH	0.44	71.2			
Ser	$HOCH_2$	0.04	54.9			
AsN	CH_2CONH_2	−0.01	72.4			
GIN	$CH_2CH_2CONH_2$	−0.10	92.7			
Asp	CH_2COOH	0.54	68.4			
Glu	CH_2CH_2COOH	0.55	84.7			
Lys	$CH_2(CH_2)NH_2$	1.50	105.1			
Arg	$(CH_2)_3NHC(NH)NH_2$	0.73	109.1			

constant and equal to 0.11 $\pm$ 0.03 kcal cm^{-3} for such "purely" hydrophobic groups; hydrogen bond forming groups lower this ratio while weak hydrogen bond forming ones, as SH and SCH_3, roughly compensate their own share to hydrophobic interaction.

The problem arises how G_X values deduced from partition studies on different groups could be applied to a quantitative characterisation of hydrophobic interactions. For groups which are neutral and do not engage in hydrogen bonding, G_X values in Tables 3.10 and 3.11 may be used as such in quantitative evaluations of hydrophobic interactions: the few $\Delta(\Delta G_X)$ values from haptene-antibody interactions agree fairly well with the G_X values mentioned above (see Table 3.10). Similarly, ΔG's for the passing of a hydrophobic molecule from water to a (relatively) nonpolar solvent, depend rather little on nonpolar solvents. The approximate constancy of G_X/R_X ratio in purely hydrophobic groups suggests the following equation for calculation of hydrophobic interaction:

$$G_X^{HP}\ (\text{kcal mol}^{-1}) = 0.11\ R_X\ (\text{cm}^3\ \text{mol}^{-1}) \tag{3.23}$$

to which hydrogen bonds, electrostatic interactions, etc. could be added. Such an equation is justified, at least partially, by the proportionality of dispersion force potential to polarisabilities (that is molar refractivities) of groups in contact (Table 3.4).

If a hydrophobic group is dissociated from a rigid hydrophobic cavity, the ΔG value may be twice that of the hydrophobic group:

$$\Delta G_{cav}^{dis} \cong 2\Delta G_X^{HP} \tag{3.23a}$$

since, unlike the passing of a hydrophobic molecule from a non-polar solvent to water, both the group and cavity come into contact with water.

More special effects in hydrophobic interactions are to be mentioned; they are present in micelle formation which is a co-operative process. From the critical concentration for self-association of cholesterol in aqueous solutions, a ΔG of —10.5 kcal mol^{-1} (25 °C) has been inferred for addition of one cholesterol molecule to the micelle [65]. Folding of long aliphatic chains lowers the transfer to water ΔG for long-chain molecules as fatty acids; the linearity of ΔG *versus* chain length holds only up the $C_{15}H_{31}COOH$ above which self association is difficult to avoid [64a]. Different $CH_3(CH_2)_n$ X molecules have a similar dependence of ΔG versus n, irrespective of n and nature of passing to hydrophobic medium (solvent, micelle, etc.); *ca.* —0.8 $\pm$ 0.1 kcal mol^{-1} is the contribution of each CH_2 group [64b, 64c]. Associations of such tensioactive long-chain molecules with proteins have a similar behaviour only up to an upper limit n since hydrophobic cavities in proteins are of limited size. The β-lactoglobulin molecule, for example, has two hydrophobic sites of about 210 $cm^3\ mol^{-1}$ each, which may accommodate a n-butyl but not a n-pentyl residue [64c]. The ionic strength of the aqueous solution also influences the hydrophobic interactions, generally reducing the solubilities of hydrophobic groups: in 2M uni-univalent saline solutions, ΔG increases with *ca.* 0.1 kcal mol^{-1} [66].

The hydrophobicity constant, π, may sometimes be used directly in structure-biological activity correlations [71]. For example, correlational equations between biological activity, A, usually defined as the logarithm of concentration (see Appendix 3), and the hydrophobicity constant π can be written in the following instances [72]:

(i) inhibition of adenosine-deaminase by 9-benzyladenines

$$A = 0.453\pi - 1.194, \quad N = 8, \quad r = 0.992 \tag{3.24a}$$

(ii) inhibition of Hill reaction in chloroplasts by some urea and amide derivatives

$$A = 0.85\pi + 2.75, \quad N = 9, \quad r = 0.920 \tag{3.24b}$$

(iii) inhibition of Hill reaction by N-phenyl-ethyl-carbamates

$$A = 0.967\pi + 3.318, \quad N = 7, \quad r = 0.985 \tag{3.24c}$$

The linear regression coefficients r and number N of compounds for which correlation equations were established (Appendix 4) are also indicated.

3.3.8. *Stacking interactions*

Stacking interactions between superposed and parallel aromatic cycles should be stronger than between aliphatic or aliphatic and aromatic groups of comparable size: the polarisability perpendicular on aromatic cycles is large and coplanarity allows atoms to come closer, which entails a strengthening of dispersion interactions (see Table 3.4). Actually, for passing of benzene molecule from water to aliphatic solvents $\Delta G \cong -0.3$ kcal mol^{-1}, and to aromatic solvents, $\Delta G \cong -1.8$ kcal mol^{-1} (at 25 °C). For a Phe-Leu pair interaction $\Delta G = -0.5$ kcal mol^{-1} while for a Phe-Phe pair $\Delta G = -1.4$ kcal mol^{-1}, according to Nemethy [19a]. The difference of -1.5 kcal mol^{-1} between the affinities of benzene for aromatic and aliphatic media may be regarded as coming from the surplus of dispersion interaction characteristic to aromatic systems. This justifies the introduction of aromatic character as a special type of intermolecular force which adds up with the ordinary hydrophobic interaction [37].

The energy of aromatic interaction — the stacking energy — is probably proportional to the surface of planar superposing cycles. Considering hexagon over pentagon surface area ratio together with a $\Delta G_{AR} = 1.5$ kcal mol^{-1} value for hexacycles (benzene, pyridine, pyrimidine) leads, for pentacycles (pyrrole, imidazole, etc.), to $\Delta G_{AR} = -1.0$ and for condensed cycles, to less than the sum of ΔG_{AR} for the component cycles.

Calculations of stacking enthalpies were carried out for superposed DNA double helix (section 3.1.1, Tables 3.2 and 3.2a). Calculations for simple homonucleotide chains (simple helix) give stronger stacking interactions in purine bases (larger areas —7 ... —9 kcal mol^{-1}) than in pyrimidine bases (smaller areas, 0 ...—4.5 kcal mol^{-1}). Calculated stacking energies between base pairs in the double helix depend on the pair of superposed bases $\left(-11 \text{ kcal mol}^{-1} \text{ or } \left\uparrow\begin{matrix}C & G\\ G & C\end{matrix}\right\downarrow \text{ and } -5 \text{ kcal mol}^{-1} \text{ for } \left\uparrow\begin{matrix}T & A\\ A & T\end{matrix}\right\downarrow\right)$. As to the experimental determinations of stacking enthalpies, ΔH (kcal mol^{-1}), for an adenine pair (AA) in simple chains, a value of 6.5 comes out [51] (calculated: —9.5, Table 3.2a) for the interaction of $\left\uparrow\begin{matrix}A & U\\ A & U\end{matrix}\right\downarrow$ base pair and —5.5 [52] (calculated: —7.4, Table 3.2) for $\left\uparrow\begin{matrix}A & T\\ A & T\end{matrix}\right\downarrow$. The stacking energies between different base pairs obtained from a semi-empirical analysis of phase diagrams are all within -5.5 ± 0.2 kcal mol^{-1} regardless of base pair $\left(\left\uparrow\begin{matrix}A & T\\ A & T\end{matrix}\right\downarrow, \left\uparrow\begin{matrix}A & T\\ T & A\end{matrix}\right\downarrow, \left\uparrow\begin{matrix}C & G\\ C & G\end{matrix}\right\downarrow, \left\uparrow\begin{matrix}A & T\\ C & G\end{matrix}\right\downarrow, \text{ and } \left\uparrow\begin{matrix}A & T\\ G & C\end{matrix}\right\downarrow\right)$ which conflicts with structural calculations (section 3.1.1, Table 3.2), strongly dependent on the interacting base pair. The dependence of stacking interactions on contact areas agrees with the findings that polyuridine exists in water as a random coil as well as depurinated DNA, while polyadenine or DNA containing purine bases as a double helix [50]. For dimerisation of nucleosides in water solution, $\Delta G = -1.5$ kcal mol^{-1} was obtained for deoxyadenosine and $\Delta G = +0.2$ kcal mol^{-1} for uridine, both dimerisations being due to stacking and not to hydrogen bonding [54].

The DNA double helix bonds easily to planar aromatic cycles by intercalation; for example dyes like acridines or ethidium bromide, bond with a rather small overwinding of the DNA helix (about 13°) and the enthalphy is of —6 ... —10 kcal mol^{-1} [69, 70]. Studies on intercalation of aromatic polycyclic hydrocarbons in the polyadenine double helix have shown that the intercalation energy is a maximum when the shape and surface area of the hydrocarbon allow maximum overlap with di-adenine pairs [48]. This maximal surface overlap rule, together with the polarities of the cycles suggest some interaction specificity of aromatic amino acid side-chains in DNA double helixes of various base compositions or in associations of aromatic amino acids with nucleosides. Thus triptophane should prefer guanine to adenine if the hexacycle stacks over hexacycle and the pentacycle moiety over the pentacycle with superposed dipole moments of opposite sense (see Figure 3.2); both should be preferred to cytidine and thymidine according to surface overlap rule. Table 3.12 lists some association constants, K_a of free amino acids in water solution [49, 67, 68]; the expected specificity seems rather low.

TABLE 3.12 **Interactions of amino acids with nucleotides**

Nucleoside / *Amino acid*	poly A	GMP	G	I	A	U	C
Trp	≅7	17.3	8.6	≅7.5	6.6	3.2	0.4
Tyr	—	7.1	—	—	—	—	—
Phe	≅2	3.7	—	—	2.9	—	—
His	≅1	—	—	—	1.4	—	—
Arg	—	—	—	—	1.5	—	—
Asp, Glu, Meth, Pro	—	—	—	—	≅0.5	—	—
Thr	—	—	—	—	≅0.15	—	—
Ser	—	—	—	—	≅0.05	—	—
Gly, Aly, Val, Leu	—	—	—	—	≅0.00	—	—

K_{as} (M^{-1}) values are listed at 25 °C; G, I, A, U, C are the respective bases.

3.3.9. *Charge-transfer complexes*

Charge-transfer complexes are formed between molecules largely differing in electronegativities and usually with planar side surfaces which allow a large contact area. The surplus of interaction energy should come from the resonance of two limit structures, AD and A^-D^+ (the last, with an electron transferred from the donor to the acceptor). The wave function which describes such complexes is a linear combination of the wave functions for the two limiting structures:

$$\psi = a\psi_{AD} + b\psi_{A^-D^+} \tag{3.25}$$

As for example the percentage $100\, b^2/(a^2 + b^2)$ of the polar limiting structure in the tetracyanoethylene-naphthalene complex should be of 6.3%, and for the I_2-pyridine complex, 25%, as indicated by dipole moment measurements [56]. As to the donor-acceptor binding energies, some ΔH values are given in Table 3.13. The first $\Delta H/2$ figures (crystalline complex as compared to components in vapour phase) represent, approximately the donor-acceptor bond energy. The ΔH figures for aliphatic non-polar solvents (CCl_4, $C_2H_2Cl_4$) represent the excess of interaction energy over the usual dispersive interactions without charge transfer. Some ΔG, ΔH and ΔS values for the building up of such complexes are listed in Table 3.13a. Per hexacycle, the excess over the usual hydrophobic interaction (which should be about the same as with carbon tetrachloride, the solvent) is *ca.* 5 kcal mol^{-1}. The contribution of aromatic character per hexacycle being *ca.* 0.75—0.9 kcal mol^{-1}, the remaining 4 kcal mol^{-1} per hexacycle should be the "pure" contribution of the charge transfer effect.

TABLE 3.13 **Heats of formation, ΔH (kcal mol^{-1}) for complexes of 1, 3, 5-trinitrobenzene with benzene, naphthalene and anthracene** [56]

Conditions		C_6H_6	$C_{10}H_8$	$C_{14}H_{10}$
$\Delta H/2$ — for crystalline complex compared to components in vapor phase		—17.5	—20.4	—23.5
$\Delta H/2$ — for crystalline complex compared to crystalline components		—0.35	—0.56	—0.17
ΔH — in solution of:	CCl_4	—1.71	—4.31	—4.5
	$C_2H_2Cl_4$	—	—	—3.6
	$C_2H_5OC_2H_5$	—	—0.99	—2.1
	C_6H_6	—	—1.90	—0.98

TABLE 3.13a **Equilibrium parameters describing formation of charge transfer complexes; 20 °C according to [56]**

Complex	*Solvent*	$K(M^{-1})$	ΔG(kcal mol^{-1})	ΔH(kcal mol^{-1})	ΔS(e.u.)
Hexamethylbenzene-chloranil	CCl_4	9.25	—1.30	—5.35	—13.8
Hexamethylbenzene-chloranil	cyclo-hexane	28.9	—1.97	—	—
Dimethylaniline-chloranil	CCl_4	3.40	—0.72	—5.05	—14.8
Hexamethylbenzene-trinitrobenzene	CCl_4	7.10	—1.15	—4.71	—12.2
Naphthaline-trinitrobenzene	CCl_4	4.00	—0.81	—4.31	—12.0
Benzene-trinitrobenzene	CCl_4	0.23	+0.84	—1.71	— 8.7
$(CH_3)_2C{=}C(CH_3)_2$-trinitrobenzene	CCl_4	0.035	+1.97	—1.46	—11.7
Benezene-I_2	CCl_4	0.15	+1.11	—1.3	— 8.2

After perturbational considerations [56], the charge transfer interaction energy, U_{CTC}, should be:

$$U_{CTC} = -\frac{W^2}{I_D - A_A - e^2/R_{AD}} \tag{3.26}$$

where W is the donor-acceptor interaction energy; I_D is the ionisation potential of the donor; A_A is the electron affinity of the acceptor and R_{AD} the donor-acceptor distance within the complex. Transposed in the language of the HMO method, formula (3.26) signifies that in order to possess a donor character, the molecule should have the highest occupied molecular orbital sufficiently high — one considers donor character for $\varepsilon_{HO} \gtrsim \alpha + 0.5\ \beta$; for acceptor character, there should be a sufficiently low empty (lowest) orbital — one considers acceptor character for $\varepsilon_{LE} \lesssim \alpha - 0.5\ \beta$ [56]. According to this criterion, from the amino acids only triptophane should present a donor character ($\varepsilon_{HO} = \alpha + 0.534\ \beta$) while from the nucleic acid bases, guanine ($\varepsilon_{HO} = \alpha + 0.307\ \beta$), adenine ($\varepsilon_{HO} = \alpha + 0.486\ \beta$) and thymine ($\varepsilon_{HO} = \alpha + 0.510\ \beta$); no member of these classes of compounds should possess electron-acceptor character. Electron-acceptor character in charge-transfer complexes may appear in some steroid rings containing carbonyl groups conjugated with double bonds (for the unconjugated $\rangle C{=}O$ group, the calculated ε_{LE} is $\alpha - 0.62\ \beta$).

3.4. The types of intermolecular forces yielded by amino acids and nucleic acid components

Table 3.14 shows schematically the types of intermolecular forces which may imply amino acid side chains, segments of polypeptidic chain, N-bases of nucleic acids, the (deoxy)ribosylphosphate and base-pairs in the shallow groove (the bottom part on drawings of base-pairs) and in the wide groove (the top part on the drawings of base-pairs) — partially according to [37]. The symbol "O" means absence of the corresponding type of intermolecular force, while "+" signifies its presence. HP stands for hydrophobic character, for a contribution (to hydrophobic force) larger than 1.5 kcal mol^{-1} ($-G_X$ from Table 3.11); the N-bases are soluble in water, and that is why there stands "O" for HP. The AR character stands for the possibility of an aromatic stacking interaction, in general for all aromatic rings. The DCT character refers to the donor character in complexes with charge transfer, while PH refers to the existence of some proton-donating (+) or proton-accepting (—) groups in hydrogen bonding; EC denotes the electrostatic charge (+ or —) in aqueous solution at pH = 7. For the side chain of histidine having an isoelectric pH at about pH = 7, both the protonated and non-protonated forms have been discussed. The sign of μ points to the existence of an appreciable dipole moment, not attached to a proton donor or acceptor group in hydrogen bonding; "←" means the negative end of the dipole directed (approximately) towards the polypeptidic chain (or towards the pentose-phosphate backbone in nucleic acids), while "→" the positive end towards the polypeptidic chain; "↑" and "↓" represent the dipole moment directed upwards and downwards respectively, against the bonding of the group to chain. For these approximate directions use has been made of Figure 3.2 (section 3.3.6). The abbreviations "sh. gr." and "w. gr." stand for the groups presented by base-pairs in the DNA double helix in the shallow groove and wide groove, respectively. In the shallow groove, the AT pair presents merely proton-accepting groups $\left(\rangle C{=}\ddot{\underset{\cdot\cdot}{O}} \text{ and } {=}\underset{\cdot\cdot}{N}{-}\right)$; in the wide groove both of the pairs present both proton donating ($-NH_2$, the free electronic pair is engaged in conjugation) and accepting $\left(\rangle C{=}\ddot{\underset{\cdot\cdot}{O}}\right.$ and $\left.-N{=}\right)$ groups. Table 3.14 may be used as a guide in looking, for a certain group, complementary groups from the viewpoint of types of intermolecular forces.

TABLE 3.14 **Types of intermolecular forces**

	Group	HF	AR	DCT	PH	EC	μ
Amino acid side chains	Gly	0	0	0	0	0	0
	Ala	0	0	0	0	0	0
	Cys	0	0	0	0	0	0
	Meth	0	0	0	0	0	0
	Val	+	0	0	0	0	0
	Leu	+	0	0	0	0	0
	Ileu	+	0	0	0	0	0
	Pro	+	0	0	0	0	0
	Phe	+	+	0	0	0	0
	Tyr	+	+	0	+	0	←
	Trp	+	+	+	+	0	→
	Ser	0	0	0	±	0	0
	Thr	0	0	0	±	0	0
	AsN	0	0	0	±	0	0
	GIN	0	0	0	±	0	0
	Asp	0	0	0	—	—	0
	Glu	0	0	0	—	—	0
	$LysH^+$	0	0	0	+	+	0
	$ArgH^+$	0	0	0	+	+	0
	$HisH^+$	0	+	0	+	+	0
	His	0	+	0	±	0	↓
Polypeptide chain	—NH—CH—CO—	0	0	0	±	0	0
	$—NH—CH—COO^-$	0	0	0	±	—	0
	$^+H_3N—CH—CO—$	0	0	0	±	+	0
	Adenine	0	+	+	±	0	→
	Guanine	0	+	+	±	0	↑
	Cytosine	0	+	0	±	0	↓
	Thymine	0	+	+	±	0	→
	Uracyl	0	+	0	±	0	→
Double helix	(Deoxy)ribosyl-phosphate	0	0	0	±	—	0
	AT (sh. gr.)	0	0	0	—	0	0
	AT (w. gr.)	0	0	0	±	0	0
	CG (sh. gr.)	0	0	0	±	0	0
	CG (w. gr.)	0	0	0	±	0	0

REFERENCES

1. J. O. Hirschfelder, *Intermolecular Forces*, McGraw Hill Book Co., New York, (1970).
2. J. Bertran, *J. Theoret. Biol.*, **34**, 353 (1972).
3. D. F. Bradley, S. Lifson and B. Honic, in: *Electronic Aspects of Biochemistry*, B. Pullman (editor), Academic Press, New York, (1964), p. 77.
4. A. Pullman and B. Pullman, *Advan. Quant. Chem.*, **4**, 267 (1968).
5. D. F. Bradley, *J. Macromol. Sci. Chem. A.*, **4**, 741 (1970).
6. A. M. Liquori, *Quart. Rev. Biophys.*, **2**, 65 (1969).
7. V. I. Poltev and B. I. Sukhorukov, *Biofizika*, **13**, 941 (1968).
8. R. Rein, J. R. Rabinowitz and T. J. Swissler, *J. Theoret. Biol.*, **34**, 215 (1972).
9. J. Mantione and B. Pullman, *Comp. Rend. Acad. Sci., Ser. D*, **262**, 149 (1966).
10. P. Claverie, *J. Molec. Biol.*, **56**, 75 (1971).
11. H. A. Scheraga, *Chem. Rev.*, **71**, 195 (1971).
12. F. K. Winkler and J. D. Dunitz, *J. Molec. Biol.*, **59**, 169 (1971).
13. W. H. Stockmayer, *J. Chem. Phys.*, **9**, 398 (1941).
14. J. L. Webb, *Enzymes and Metabolic Inhibitors*, Academic Press, New York, (1963) Ch. 6.
15. A. Cammarata, in: *Mol. Orbital Study Chem. Pharmacol. Symp.*, L. B. Kier (editor), Springer, New York, (1970), p. 156.
16. G. Kloppman and R. F. Hudson, *Theoret. Chim. Acta*, **8**, 165 (1967); G. Kloppman, *J. Amer. Chem. Soc.*, **90**, 223 (1968).
17. G. J. Hoijtink, E. De Boer, D. M. van der Meij and W. P. Weijland, *Rec. Trav. Chim.*, **75**, 487 (1956); G. Kloppman, *Chem. Phys. Letters*, **1**, 5 (1967).
18. J. D. Bernal and R. W. Fowler, *J. Chem. Phys.*, **1**, 315 (1933).
19. G. Némethy and H. A. Scheraga, *J. Chem. Phys.*, **36**, 3382, 3401 (1962); *J. Phys. Chem.*, **66**, 1773 (1962).

19a. G. Némethy, *Angew. Chem.*, **79**, 2608 (1967).

20. O. Y. Samoilov, *Discuss. Faraday Soc.*, **24**, 141 (1957).
21. K. D. Gibson and H. A. Scheraga, *Proc. Nat. Acad. Sci. U.S.A.*, **58**, 420 (1967).
22. G. Schwarzenbach, *Z. Physik. Chem.*, **176**, 133 (1936).
23. A. D. Buckingham, *Quart. Revs. Chem. Soc.*, **13**, 183 (1959).
24. E. Chargaff and J. N. Davidson, *The Nucleic Acids*, Academic Press, New York, (1955) vol. I, p. 465.
25. *Handbook of Chemistry and Physics*, The Chemical Rubber Co., Cleveland, Ohio, (1955—1956) vol. II, p. 1649.
26. D. Pressman and M. Siegel, *J. Amer. Chem. Soc.*, **75**, 686 (1953).
27. A. Nisonoff and D. Pressman, *J. Amer. Chem. Soc.*, **79**, 1616 (1957).
28. F. Bergman, *Advan. Catalysis*, **10**, 130 (1958).
29. S. Lewis, *J. Theoret. Biol.*, **23**, 279 (1969).
30. J. M. Lehn, *Angew. Chem.*, **82**, 183 (1970).

31. B. D. Lindley, *J. Theoret. Biol.*, **17**, 213 (1967).
32. L. N. Ferguson, *The Modern Structural Chemistry*, Prentice-Hall, Inc., Englewood Cliffs, N. Y., (1964).
33. G. C. Pimentel and A. C. McClellan, *The Hydrogen Bond*, W. H. Freeman & Co, San Francisco and London, (1960).
34. I. M. Klotz and J. S. Franzen, *J. Amer. Chem. Soc.*, **84**, 3461 (1962).
35. A. J. Kipper, L. V. Dmitrenko, O. B. Ptitsyn and J. S. Bogomyantz, *Molek. Biol.*, **4**, 175 (1970).
36. M. Laskowski, jr. and H. A. Scheraga, *J. Amer. Chem. Soc.*, **76**, 6305 (1954).
37. Z. Simon, *Rev. Roum. Biochim.*, **5**, 319 (1968).
38. F. S. Binford, jr. and D. M. Holloway, *J. Molec. Biol.*, **31**, 91 (1968).
39. P. B. Ptitsyn, *Uspekhi Sovrem. Biol. (Adv. Mod. Biol.)*, **63**, 3 (1967).
40. N. I. Nakano and S. J. Igarashi, *Biochemistry*, **9**, 577 (1970).
41. H. A. Scheraga, *Ber. Bunsenges. Phys. Chem.*, **68**, 838 (1964).
42. A. Ray, *Nature*, **231**, 313 (1971).
43. A. Leo, C. Hansch and D. Elkins, *Chem. Revs.*, **71**, 525 (1971).
44. Ch. Tanford, *J. Amer. Chem. Soc.*, **84**, 4240 (1962).
45. J. M. Zimmerman, N. Eliezer and R. Sinha, *J. Theoret. Biol.*, **21**, 170 (1968).
46. P. Cornea, Licenciate's Work, University of Timişoara, (1969).
47. Ch. C. Bigelow, *J. Theoret. Biol.*, **16**, 187 (1967).
48. A. M. Craig and I. Isenberg, *Proc. Nat. Acad. Sci U.S.A.*, **67**, 1337 (1970).
49. M. Razska and M. Mandel, *Proc. Nat. Acad. Sci. U.S.A.*, **68**, 1190, (1971).
50. Achter and Felsenfeld, *Biopolymers*, **10**, 1625 (1971), *Nature*, **234**, 128 (1971).
51. D. Poland, N. J. Vournakis and H. A. Scheraga, *Biopolymers*, **4**, 223 (1966).
52. W. H. Huang and P. O. P. Ts'O, *J. Molec. Biol.*, **16**, 523 (1966).
53. N. S. Goel, N. Fukuda and R. Rein, *J. Theoret. Biol.*, **18**, 350, (1968).
54. T. N. Solie and J. A. Schellman, *J. Molec. Biol.*, **33**, 61 (1968).
55. P. Spacu and M. Brezeanu, *The Chemistry of Complex Compounds* (in Romanian), Publishing House for Didactics and Pedagogy, Bucharest, (1969) Chs. 11, 12.
56. G. Briegleb, *Electron Donating-Accepting Complexes* (in German), Springer Verlag, Berlin, (1961) Chs. III, IX.
57. A. Szent-Györgyi, *Introduction to a Submolecular Biology*, Academic Press, London, (1960) pp. 38—41.
58. D. Pressman, in: *Molecular Structure and Biological Specificity*, L. Pauling and H. A. Itano (editors), Amer. Inst. Biological Sci., Washington, (1957) p. 1.
59. D. R. Williams, *Chem. Rev.*, **72**, 203 (1972).
60. S. L. Brenner and D. A. McQuarrie, *J. Theoret. Biol.*, **39**, 343 (1973).
61. A. Katchalsky, *Pure and Appl. Chem.*, **26**, 327 (1971).
62. A. R. Fersht, *J. Molec. Biol.*, **64**, 497 (1972).
63. I. V. Zuika and Iu. A. Bankovskii, *Usp. Himii*, **42**, 39 (1973).

64(a). R. Smith and Ch. Tanford, *Proc. Nat. Acad. Sci. U.S.A.*, **70,** 289 (1973); (b). Ch. Tanford, *J. Molec. Biol.*, **67,** 59 (1972); c. R. Smith and Ch. Tanford, *idem*, 75.

65. M. E. Haberland and J. A. Reynolds, *Proc. Nat. Acad. Sci. U.S.A.*, **70**, 2313 (1973).

66. P. K. Nandi and D. R. Robinson, *J. Amer. Chem. Soc.*, **94**, 1299, 1308 (1972).
67. V. I. Bruskov and A. I. Klimov, *Biofizika* (*U.S.S.R.*), **17**, 151 (1972).
68. H. A. Aarfmann and K. F. Wagner, *Commun. at the VIth Jena Molecular Biology Symposium*, (1973); N. Saringa, *Nature New Biol.*, **234,** 172 (1970).
69. J. Paoletti and J. B. LePecq *J. Molec. Biol.*, **59,** 43 (1971).
70. P. H. von Hippel and J. M. Mc. Ghee, *Ann. Rev. Biochem.*, **41,** 790 (1972).
71. I. Schwartz, *Studii Cercet. Chimie*, **21,** 721 (1973).
72(a). H. J. Schaffer, R. N. Johnson and E. Oddin, *J. Med. Chem.*, **13,** 452 (1970); (b) C. Hansch and E. W. Deutsch, *Biochim. Biophys. Acta*, **112,** 381 (1966).

4. Electronic structure and reactivity of biomolecules

This chapter deals with the connection between electronic structure and reactivity of biomolecules, their absorption and emission spectra, as well as other physical properties. The compounds dealt with in this chapter are: proteins and nucleic acids with the corresponding monomers; there will also be discussion on the theories of enzymatic activity, as well as on some classes of molecules with important biological activity. Various quantum chemical calculations, especially those regarding reactivity or directly biological activity of substances quoted above will be also reviewed.

4.1. *Proteins and amino acids*

The amino acids (Figure 4.1) do not present large systems of conjugated bonds; however, there are only four aromatic amino acids. Therefore, Hückel's [1] calculations have rapidly dealt with all amino acids, and the calculations involving all the valence electrons have been carried out with the special aim of the energy study of various possible spatial configurations. The proteins, especially in crystalline forms, present small conjugated systems, which are relatively weakly bonded to one another — the peptide groups bonded to each other by hydrogen bonds (Figure 4.2). The >CHR groups with C atoms in sp^3 hybridisation interrupt a direct conjugation between the consecutive —CONH— groups on the polypeptide chain. For this reason the relatively distant levels of the —CONH— groups are not shifted too much by the formation of these hydrogen bonds; however, rather narrow energy bands arise throughout the crystalline portions of proteins [1]. These energy bands, together with the impurities and the polarisation effects of the hydration water, are responsible for the semiconductor properties found in proteins [2].

The spectral properties of proteins — absorption bands are not too far in the ultraviolet range (240—300 mμ) — are due to aromatic amino acids. The aromatic amino acids constitute too little of the proteins to

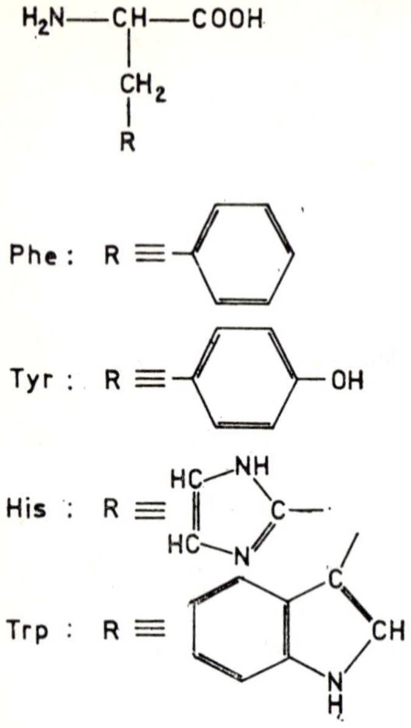

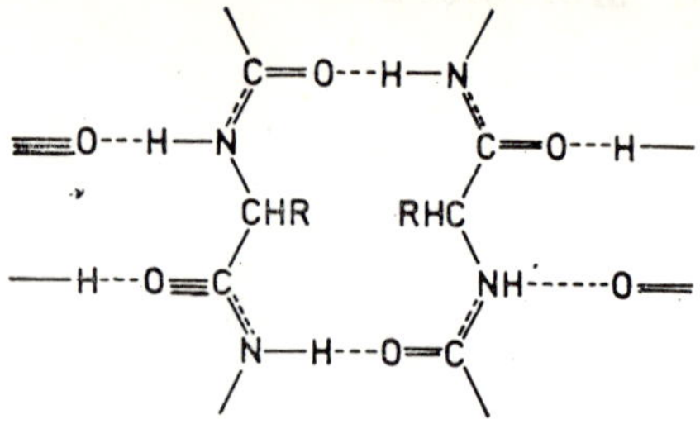

Figure **4.2.** Formation of extended quasi-conjugated systems by means of hydrogen bonding in polypeptides.

Figure **4.1.** The aromatic amino acids

enable us to speak about energy bands originating in the interaction of aromatic rings. The transition of electronic excitation energy between these aromatic residues and, eventually, from the excited peptide band of these amino acids, is a frequent phenomenon. It is this way that one explains the luminiscence properties of the proteins, the emission occurring from the excited levels of tyrosine and, especially, of tryptophan: the latter presents, through its indole ring, the lowest energy levels in the protein molecule, acting as a "trap" for the electronic excitation energy [3].

Concerning the reactivity of amino acids in proteins, it should not differ from that, fairly common, of the side chains and of the corresponding peptide groups. However, a very high importance is attached to the spatial structure of proteins, which holds the various groups in relatively rigid positions, allowing simultaneous reactions and interactions between these groups and molecules of a certain form, while sterically screening, on the other hand, other amino acid groups.

4.1.1. *Amino acids*

HMO calculations were performed for aromatic side chains of natural amino acids, and are reviewed in the work "Quantum Biochemistry" of Pullman and Pullman ([1], Table 49). Some of the HMO results are mentioned in Figure 3.2 (chap. 3) and sections 3.3.8 and 3.3.9.

From the more advanced calculations we may mention the HMO + Del Re calculations of Berthod and Pullman [64] for charge distributions and dipole moments in indole (Trp) —2.05 D; and imidazole (His)—4.03 D; the experimental figures are 2.05 D and 3.99 D respectively. The protonation of the peptidic CONH unit was studied by Moffat [65] using a non-empiric LCAO—MO—SCF method with gaussian atomic unctions. The computed potential surfaces indicate that the —NH— group is protonated, but if the rotation of the obtained —NH_2^+— group is not permitted, the protonation of the O-atom is favoured. Finally, one should mention a rather recent review of Alberte Pullman about advanced calculations at all or quasi-all valence electron levels, for amino acids, dipeptides, constituents of nulcleic acids, etc. [66].

4.1.2. *Polypeptides and proteins*

Evans and Gergely [4] have carried out calculations of energy levels in infinite chains of peptide groups bonded by hydrogen bonds, by the HMO method; Suard, Berthier and Pullman [5] have compared calculations for the groups of peptides, dipeptides and polypeptides by means of a self-consistent MO method, only for π- and *n*-electrons. Within the π-electron approximation, the peptide group shows up with three molecular orbitals occupied by 4π-electrons, and additionally, a free electron pair (not involved in hydrogen bonding) at oxygen. According to calculations of Suard, Berthier and Pullman [5], the two π molecular orbitals occupied by electrons are located at —15.04 eV and —12.68 eV, the n orbital at —12.63 eV, and the free antibonding orbital at +1.24 eV. The energy of the nπ * singlet (5.8 eV) electron transitions is in agreement with the wave length of the first two bands in the electronic spectra of the peptide group (1900—2500 Å, i.e. 5—6.5 eV, and 1717 Å, 7.2 eV). For polypeptide, energy bands corresponding to these molecular orbitals, a width of about 1.2 eV are obtained; the ionisation potential decreases by 0.9 eV, while the electronic affinity increases by 0.6 eV in comparison to the monopeptide. The calculated energy of the first $n\pi^*$ transition for a polypeptide is about 5 eV, and the difference between the upper limit of the last occupied band and the lower limit of the first empty band of 12.5 eV. These data calculated for a polypeptide must be compared to the width of the forbidden band, $\Delta\varepsilon$, calculated from the temperature dependence of the semiconductivity of some crystalline proteins, according to:

$$R = R_o \, e^{-\Delta\varepsilon/2kT} \tag{4.1}$$

where R and R_0 are the resistivity at temperature T and extrapolated to 0 °K. The study of resistivity for various proteins in crystalline-state gives for $\Delta\varepsilon$ values in the 2.5—3 eV range [6].

The resistivity is much lowered by impurities and especially by hydration, in which case the conductivity may be due to a transfer of protons above the hydrogen bonds [7]. These semiconductor properties of proteins might be of no importance in their physiological activity, since in physiological conditions we come across aqueous solutions or at least highly hydrated proteins. The enzymes involved in redox redactions, contain prosthetic groups — large conjugated rings, eventually transition metal ions of variable valence — which can, by themselves, account for the electron donor and acceptor properties of these enzymes.

More advanced calculations by MO methods in a selfconsistent field and with consideration of all valence electrons, have been carried out for polypeptide molecules by Imamura *et al.* [8], and for a series of amino acid residues — like those found in proteins, as dipeptides or as homopolypeptides — by Pullman *et al.* [9]. Imamura's calculation regards the electron spectra, the ionisation potential and the dipole moment of the peptide group; the computed values are in good agreement with experimental results. In addition to simpler calculations only for π-electrons, Imamura's calculation reveals that in electron spectra of the peptide group the $\pi\pi^*$-transitions might be "mixed" — through configurational interaction — with $\sigma\sigma^*$, $\pi\sigma^*$ and $\sigma\pi^*$-transitions. The same calculations point to an important role of the imidic proton in protein semiconductivity. The calculations of Pullman *et al.* for all the valence electrons aim at clearing up the energetics of spatial configuration. The energy minima predicted by these calculations are obtained for spatial configurations which are close to those predicted by empirical calculations; these latter are based upon the energies calculated as sums of energies of repulsion between pairs of atoms, sums of harmonic potentials for angular distortions, torsions, etc. For example, Pullman's calculation for diproline shows the *trans* conformer to be more stable than the *cis* against the peptide bond —with 0.5 kcal mol^{-1}, and the energy barrier to their interconversion is *ca.* 16.2 kcal mol^{-1}. That same calculation for the homopolymer of proline indicates the higher stability of the *trans* conformer by 1.4 kcal mol^{-1} compared to the *cis*.

Thus, the amide-group should be only slightly rigid against twisting around the C—N bond (partially double by conjugation). This is confirmed by X-ray studies [10] of the crystal structure of some medium-sized lactam rings, indicating a value of 10 kcal mol^{-1} for the energy barrier corresponding to this twisting [10].

4.2. Nucleic acids and bases

The five native bases in nucleic acids, as well as a series of their derivatives — important for intermediary metabolism, mutations or as anti-metabolite drugs — form a set of conjugated heterocycles i.e. aromatic systems. So, shortly after the elucidating of nucleic acids structure and their central biological rôle, a great deal of calculations of HMO type have been carried out for these bases [1a]. The numbering of atoms within these purine and pyrimidine bases is shown in Figure 4.3. For substitution derivatives the number of substituent carrying atoms is indicated too.

The ordered arrangement of these bases in the DNA double helix parallel base-pairs, nearly overlapping at a distance of 3.4 Å above one another, brings about the existence of some systems of fairly narrow bands. Thus, some spectral (optical) characteristics of bases [11], are altered and semiconductivity [12], luminiscence and energy migration [13] properties arise. A lot of the biological properties of DNA, e.g. even the control of chromosome replication or the point mutations, were assigned to these semiconductor properties and to proton tunnelling among the hydrogen bonded bases [14].

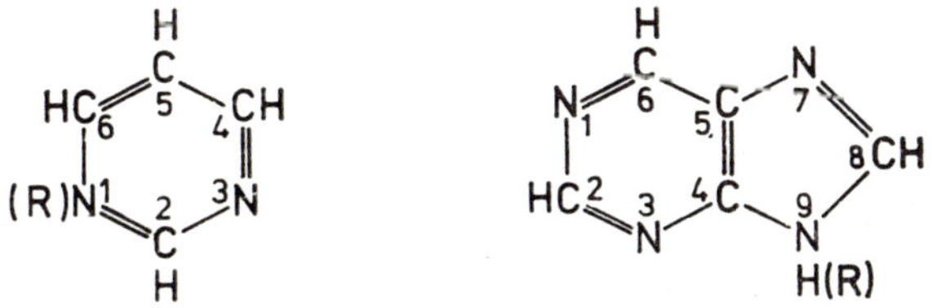

Figure **4.3.** Numbering of atoms in purine and pyrimidine; R-pentose residue.

4.2.1. *The bases*

The bases have been the object of many calculations, especially those of Pullman's group, both by the simple HMO method [16] and advanced methods [1c]. A presentation of some calculations and results may be found in Chapter 2 (paragraphs 2.2, 2.3, 2.5 and 2.7; Figures 2.3, 2.6, 2.8 and 2.9; Table 2.5). From among the advanced method calculations concerning bases, we mention those of Fraga and Carbo [15]. In Pullman's *Quantum Biochemistry* [1] are correlated several physico-chemical and reactivity characteristics of these bases with the parameters of the HMO calculation: tautomeric transformations, resonance energies, charge-transfer complex formation, basicity of bases and various atoms therein, antitumor activity of some purine antimetabolites, reactions at the amino-groups, mechanisms of enzymic cleavage (*via* xantine oxidase) of purines and their relative resistance to ionising radiations. The agreement between the order predicted by HMO parameters for various characteristic values and the one resulting from experience, is sometimes good, sometimes rather poor.

For example, for the basicity of heterocycles, the pK-values should parallel the π-electron charges, ρ_p^π, of the most negative N-atoms of the pyrimidine ring; however this is not found. Nevertheless, a good correlation is obtained by means of a formula advanced by Nakajima and Pullman [16], which takes into account the interaction between the free pair of electrons of the various N-atoms and the π-electron charges, ρ_p^π, of all the atoms in the heterocycle:

$$\mathrm{p}K = B + \sum_p \rho_p^\pi \gamma_{np} \tag{4.2}$$

in which B is a constant and γ_{np} the interelectronic repulsion integral between the free pair n and the π-electron charge on the atom p (*cf.* section 2.6). This formula predicts correctly the N-atom being more basic, and — as follows from Table 4.1 — it also predicts fairly well the relative basicity of various heterocycles. In Table 4.1, ρ_q^π is the π-electron charge of the most basic atom of the corresponding heterocycle, calculated by means of the HMO method ([1], Appendix); ρ_q^{total} is the total charge of that same atom, calculated by Fraga and Carbo [15]. As follows from Table 4.1, the order of decreasing basicity does not agree with the order of decreasing charge of the most negative N-atom, no matter what the method used for the calculation of this charge (HMO or AVE—CI—SCF).

There also exists a rather good correlation between the antimetabolite (cytostatic) activity of derivatives of native purine bases and the basicity of the N-atoms [1a]. The basicity of the active analogues is closer to that of the native bases than to the basicity of the inactive ones, whose basicity is lower. Also, the charge of the iminic N-atom, N(9), where the

TABLE 4.1 **Basicity of some purine bases**

Base	pK*a*	*n-position of most basic N atom*	$\Sigma \rho_p^{\pi} \gamma_{np}$	ρ_q^{π}	total ρ_q
2,6-diaminopurine	5.1	N(1)	—2.21	—0.357	—0.422
adenine	4.2	N(1)	—1.91	—0.305	—0.404
guanine	3.3	N(7)	—1.67	—0.422	—0.346
8-aza-adenine	2.6	N(1)	—1.65	—0.282	—0.410
6-methylpurine	2.6	N(1)	—1.40	—0.311	—0.354
6-mercaptopurine	2.5	N(7)	—1.42	—0.314	—
purine	2.4	N(1)	—1.52	—0.308	—0.349
8-azapurine	2.12	N(1)	—1.22	—0.219	—0.351
8-azaguanine	1.1	N(7)	—1.46	—0.423	—0.406
Linear correlation coefficient, *r*			0.88	0.31	0.09

glycoside-bond (with pentose) is formed, is *ca.* +0.42 for adenine and guanine, +0.40 ... +0.43 for the active analogues, and *ca.* +0.44 for the inactive analogues.

Another problem is that of the amino-imino and keto-enol tautomerism of bases: adenine and tautomeric thymine have hydrogen bonding properties similar to guanine and "normal" cytosine, respectively, while guanine and tautomeric cytosine are similar to adenine and "normal" thymine (Fig. 4.4). As regards the experimental studies on free bases in dimethyl sulphoxide solution (n.m.r. spectra), the imine tautomeric forms do not seem to exist, yet cytosine does readily transform into an amphoteric ionic enol tautomer [17]. The spectra of aqueous solutions of cytidine also indicates a prevalence of the keto-form [18].

A study of keto-enol tautomerism in aqueous solution, by comparing the base tautomerism with CH_3-substituted derivatives in positions preventing tautomerism, indicates, for keto-enol tautomerism in inosine, an equilibrium constant $K \cong 1$. With amino-imine tautomerism of adenosine, $K = 4 \times 10^4$, favouring the amino-form [18a]. A calculation of the energy difference between "normal" and tautomeric forms, based on bond and resonance energies calculated by means of the HMO method, comes up with the following results. From the viewpoint of the sum of bond energies, the amino-forms and the imino-tautomers do not differ; the lactam (keto-imine) forms are 10 kcal mol^{-1} more stable than the lactim (enol-pyridine) ones. The resonance energies favour the amino-

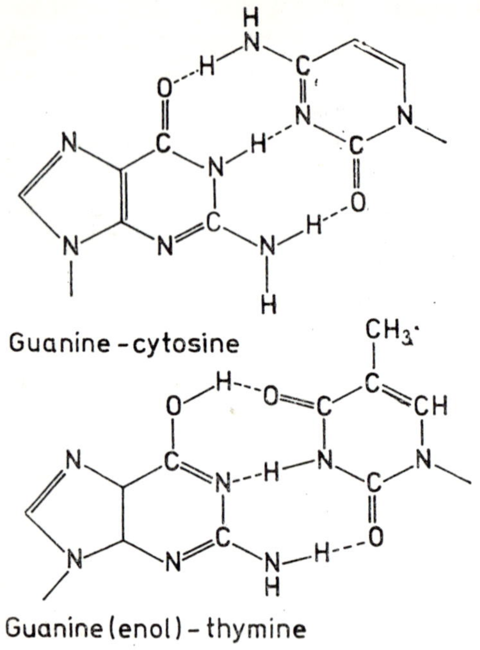

Guanine - cytosine

Guanine (enol) - thymine

Figure 4.4. "Correct" and "false" base pairing through hydrogen bonding by normal and enol tautomeric guanine, respectively.

against the imino-forms with *ca.* 0.13 $\beta = -2.1$ kcal mol^{-1} (cytosine) and the lactim against lactam forms with $0.33\beta \cong -5.5$ kcal mol^{-1} (guanine). The energy difference between the natural and "tautomeric" forms should thus be small, not exceeding 5 kcal mol^{-1}. Advanced calculations, by CNDO/2 method, indicate even the lactim form of guanine, as having 13 kcal mol^{-1} more stability than the "normal" lactam form; the "normal" forms of uracil, cytosine and adenine appear more stable than the tautomeric ones by 10.4, 26.5 and 49 kcal mol^{-1}, respectively [15]. However, the latter results are based on the difference between the total energies (very high ones) of the "normal" and tautomeric forms; very likely, these differences do not exceed the error inherent to the calculation of total energy (here: the atomising energy). Other calculations [20] take into account the energies of interaction between the atoms of bases of hydrogen bonded pairs (electrostatic interactions also involving hydrogen bonds), induction and dispersion interactions (*cf.* section 3.1.1). According to these calculations, replacing a normal AT-pair with the pairs A (imine) C; AC (imine); GT (enol) and G (enol) T, brings about an energy loss of 11.04, 0.93, 11,5 and 3 kcal mol^{-1}, respectively, while replacing GC with A (imine) C; AC (imine); G (enol) T results in a loss of 0.3, 10.4 and

8.36 kcal mol^{-1}, respectively. As a conclusion, both the results of these calculations and the existing experimental results (*cf.* also [21]) indicate a rather small loss of energy on passing from "normal" forms to some "tautomeric" ones, as well as the forming of false base-pairing with these tautomeric forms. According to these calculations [20] the probability of false base-pairing may reach fairly high values, of up to 10^{-4}—10^{-5}, or even higher.

4.2.2. *Proton tunnelling and spontaneous mutations*

DNA replication implies the temporary uncoiling of the double helix, with the synthesis of the new complementary strands on the matrix chains; in the uncoiled region the specificity of copying is ensured by the bonding — *via* hydrogen bonds between complementary bases — of a base in the matrix with a base from among the deoxynucleoside triphosphates (the precursors of the newly synthesised bases). Thus, the probability of a spontaneous fault is equal to the probability of false pairing — *via* hydrogen bonding — either by an "unusual" tautomeric base — as above — or by forming of fewer hydrogen bonds. These possibilities would yield for the probability of a spontaneous mutation, a value $p = 10^{-4}$, by far higher than $p = 10^{-8}$—10^{-11}, deduced from the natural frequency of spontaneous mutations [22].

If the false pairing between purine and pyrimidine bases in "normal" form can yield very small probabilities p, by sterical repulsion between hydrogen bonded groups, the not too low proportion of some "unusual" tautomers (in equilibrium with the 'normal' ones) requires that throughout DNA replication the base tautomerisation towards "unusual" forms would not proceed the same way with free bases in solution. Indeed, the probability of false base-pairing by an 'unusual' tautomer, will not be given by the free enthalpy loss ΔG on tautomerisation of a free base, but by the loss ΔG on the "unusual" tautomerisation of a base pair, especially if the DNA template is interacting with already hydrogen bonded deoxynucleosidetriphosphates and not with free ones [23]. A group of workers at Uppsala have carried out a series of calculations of the energy of a base-pair, as a function of proton position, in order to calculate the probability of formation of some "unusual" tautomers in a base pair by taking into account also the possibility of proton tunnelling (see [24] for a review).

This kind of calculations have been performed, e.g. for proton transfer from the "normal" form to the "unusual" one (Figure 4.5), by computing the energy of the base-pair for the two proton configurations and for a number of intermediary configurations. The first calculations were made by starting with atomic charges and bond orders (HMO method) by evaluating the potential for the transfer of the intermediate proton of GC-pair from guaninic NH to the pyridinic N of cytosine [23]. With the same parameters for base-pair, but obtained through a semi-empirical self-consistent method [25], Ladik [26] has calculated the potential "sensed" by the proton of the >N ... H—N< bond of the GC-pair, assuming that the π- and σ-electrons are not polarised by the displacement of this proton. The calculated energy difference between the two minima is *ca.* 2.72 eV (in favour of the 'normal' base), and the potential barrier is 3.40 eV. In order to consider the polarising effect of the proton, Rein and Harris [27] have developed an improved self-consistent method which explicitly

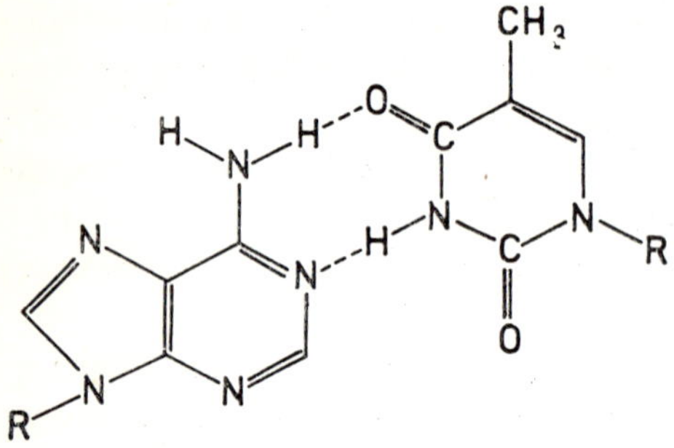

„Normal" AT-pair

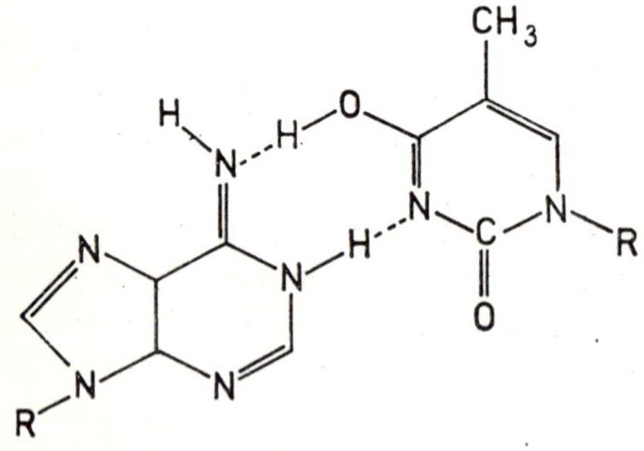

„Doubly tautomeric" AT-pair

Figure 4.5. The "normal" and "unusual" (double tautomeric) forms of adenine-thymine pair. (The resemblance of unusual AT-pair to "normal" GC-pair is remarkable).

considers the 24 π-electrons and 4 σ-electrons of the hydrogen bond. The self-consistent calculations were carried out on 12 subsequent positions of the proton along the hydrogen bond. This time, the energy difference between the minima is only *ca.* 1.4 eV, the potential barrier is 1.8 eV. Considering the simultaneous shifting of all the protons in hydrogen bonds, the resulting energy difference between minima are *ca.* 0.65 eV for the CG-pair, and *ca.* 0.70 eV for the AT-pair, in both cases in favour of "normal" forms.

In parallel with the above, calculations have also been carried out of the rate of proton transfer from the minimum corresponding to the "normal" form to that of the "tautomeric" one. This transfer takes place through a tunnelling process (the relatively easy transfer of protons through — and not over — the potential barrier) and it requires a time span of 10^{-4}—10^{-5} s, short in comparison to the one required by the replication of a base-pair (at least 10^{-3} s). Thus the probability of forming "tautomeric" forms is given by the difference of potential minima (ΔU), corresponding at 37 °C, to $p = 3\times10^{-11}$ for CG-pair, and $p = 4\times10^{-12}$ for AT, values comparable to those required by experimental data: $p = 10^{-8}$ — 10^{-11}.

An interesting result of these calculations is that in the excited electronic states, especially in charge-transfer states, which may be reached by relatively large photons (relatively far U.V. region), the difference between potential minima decreases, that is p increases. This effect may explain the mutagenic effect of U.V. radiation [24].

4.2.3. *The DNA double helix*

The DNA double helix presents several new properties — besides those of isolated bases or nucleotides — which may be explained by means of quantum-chemical calculations. Thus, there is a decrease in the absorption intensity in the U.V. spectra (hypsochromic effect), to only 60% of the sum of intensities of free bases [11]. DNA also presents semiconductor properties with a forbidden band $\Delta\varepsilon$ of 2.42 eV for perfectly dried DNA and of 1.8 eV for hydrated DNA [12, 28]. The luminiscence spectra of nucleic acids also differ from those of the isolated bases; the triplet state which appears in such circumstances (various excitations) behaves as an exciton, with an emission characteristic to purines and not to pyrimidines.

The first quantum-chemical calculations for oligonucleotides (short segments of double helix) were done by Pullman's group [1] by a HMO method. The X — H ···: Y hydrogen bonds are either approached with a $\beta_{XY} = 0.2\,\beta$ integral between atoms X and Y or the proton is treated as a separate centre, with a high $2p_z$ orbital. An outcome of these calculations is the decrease in electronic excitation energy (of the $\varepsilon_{LE} - \varepsilon_{HO}$ difference) with 0.07 β in AT against adenine, and with 0.24 β for GC against cytosine; the latter effect, strongly bathochrome, is not verified experimentally. The interaction between superposed bases, within the HMO method, is considered by introducing some β_{ij} integrals between superposed atoms, proportional to the overlap integrals between the corresponding $2p_z$ orbitals [29]. In this case too, the calculus predicts a decrease in excitation energy comparative to that of the base of lowest excitation energy in the oligonucleotide: up to 0.20 β. There also results some decrease in the oscillator strength for the first absorption band (calculated $f = 0.504$ for the superposed pairs of C—C compared to $f = 0.57$ for cytosine), which explains, at least partially, the hypsochrome effect for the transfer from free bases into the DNA double helix. Calculations were also carried out for homopolynucleotides by the same method and making use of some results of solid state theory in solving secular equations of infinite order. However, this method does not suit heteropolynucleotides. As a result, the energy bands obtained are 0.02—0.10 β wide and the excitation energy of the first electronic transition decreases by 0.05—0.09 β. A value of $\beta = -3.33$ eV (calibrated from electron spectra) leads to a width of the forbidden zone of $\Delta\varepsilon = 3.52$ eV, or — if one includes in calculation the hydrogen atoms as separated centres — $\Delta\varepsilon = 2.69$ eV. The latter calculated value is in fairly good agreement with the aforementioned experimental value. Another simple calculation [12] for DNA indicates $\Delta\varepsilon = 1.5$ eV for a strong conjugation between the bases from the superposed pairs, and in case this conjugation is absent, $\Delta\varepsilon \cong 3$ eV. The comparison between the values so calculated and the experimental ones cited previously [28] points to a weak conjugation between paired bases and to the conductivity phenomena in DNA being due to electron migration along the double-helix axis.

Calculations of the structure of energy bands in the five homopolynucleotides (poly-A, poly-T, poly-G, poly-C and poly-U) have also been carried out by a semi-empirical self-consistent method, namely the **PPP**-method, [30]. (*cf.* section 2.6). According to this calculation, the energy bands are found close to the corresponding energy levels of the free base;

the band width is smaller than the one given by the HMO method, and the forbidden zones are much wider. The energy required for promoting an electron on the conduction band comes out to be about equal to that necessary for the excitation of free bases ($\Delta\varepsilon \cong 4$—$5.5$ eV). By the same method was also calculated the structure of energy bands for periodical heteropolymers (of the poly-AT, poly-CG type), obtaining narrower energy bands (*ca.* 0.03 eV [31]). These calculations, which basically only take into account the superposed bases, yield $\Delta\varepsilon$ values far higher than those deduced from experimental measurements of semiconductivity. However the DNA samples were DNA salts (not free DNA) which generally contain hydration water too. Therefore, the PPP method with "crystal orbitals" has also been applied to models involving hydration H_2O or Mg^{2+} at different positions [32]. The presence of Mg^{2+} strongly modifies the band structure of the polymers (poly-GC or poly-AT). Thus, for the non-perturbed poly-GC pair $\Delta\varepsilon = 5.98$ eV, for hydrated poly-GC $\Delta\varepsilon = 6.12$ eV, and if the free electron pair of the amino-group of cytosine is withdrawn from the conjugated system by bonding of Mg^{2+}, $\Delta\varepsilon$ decreases to 2.00 eV. Similarly, for poly-AT, $\Delta\varepsilon$ decreases from 6.47 to 3.97 eV, following the same bonding of Mg^{2+} to the carbonylic oxygen of C_4 on thymine. These calculations, more accurate than the HMO ones, indicate an important effect of metal cations in favouring DNA semiconductivity.

4.3. Enzymic reactions and coenzymes (co-factors)

Enzyme-catalysed reactions, unlike those catalysed by usual catalysts (solid-liquid interface, solutions) possess a higher specificity regarding the substrate that is being transformed by them; they generally bring about an acceleration of some orders of magnitude higher than the non-enzymic catalysts, and besides the competitive inhibition phenomenon (by end-products or false reactants) they also present allosteric non-competitive inhibition and activation. These peculiar characteristics of the enzymic reactions are due to the large molecular size of enzymes, much larger than that of the ions or the constellation of active centres from non-enzymic catalysts. The enzymes are proteins: they have relatively rigid loci which only attach molecules of a certain shape. As often happens, the proper catalyst is a co-factor, a transition ion or relatively large heterocycle, with an extended conjugated system. The high efficiency in speeding up reactions is explained by the concerted intervention of several catalytic effects of the usual type; some authors, however, consider these effects as insufficient and look for more unusual explanations. Beyond these reaction loci or sites, the enzymes offer additional sites, to which the modulators (activators or inhibitors) may attach: the bonding to the "modulator" site does modify the spatial configuration of a portion of protein which also comprises the reaction site [33, 34].

The enzymes involved in redox processes (electron transport) and in the processes of the addition of a carbon atom (methylation, formylation) have usually, as an active part, a sufficiently large conjugated system, which is a condensed heterocycle. The kinetic characteristics of these reactions (relative easiness of attack at various positions of substrate molecule or at different substrate molecules) have been correlated to molecular orbital parameters. As co-factors are large systems, calculations especially of HMO type [16] have been carried, out but there are also some advanced calculations which consider all valence electrons [35].

4.3.1. *Kinetics of enzymatic reactions.*

Given the non-catalytic reaction

$$\mathrm{A} + \mathrm{B} \rightarrow \mathrm{P} \tag{4.3}$$

with a rate constant k_0, the corresponding reaction involving the enzyme E may be written, usually, as [34]:

$\mathrm{A} + \mathrm{E} \rightarrow \mathrm{E}'$ (fast process, moving towards the formation of E′ complex) (4.4a)

$$\mathrm{E}' + \mathrm{B} \underset{k_{-1}}{\overset{k_1}{\rightleftarrows}} \mathrm{C}$$

$$\mathrm{C} \xrightarrow{k_2} \mathrm{P}. \tag{4.4b}$$

The rate $v_0 = \mathrm{d}P/\mathrm{d}t$ of the non-catalytic reaction and the rate $v_{\mathrm{E}} = \mathrm{d}P/\mathrm{d}t$ of the enzymic reaction can be written (the small letters stand for the corresponding concentrations, A and B represent reactants; P, reaction products; E, enzyme; E′, the complex of enzyme with molecule A; C, the complex of enzyme with molecules A and B; k_1, k_{-1} are rate constants for the respective reactions) as:

$$v_0 = k_0 ab \tag{4.5a}$$

$$v_{\mathrm{E}} = \frac{k_2 eb}{b + \dfrac{k_2 + k_{-1}}{k_1}} \tag{4.5b}$$

For large enough concentrations of substrate b:

$$v_E = k_2 e \tag{4.5c}$$

$$\frac{v_E}{v_0} \cong \frac{k_2 e}{k_0 ab} \tag{4.5d}$$

Now let us find out of what constitutes the catalytic effect of enzyme. A first effect consists in achieving a very high "local" concentration of reactants, as well as a mutual orientation favourable to the reaction. Thus, the number of collisions experienced by a molecule (for 1 mol l^{-1} concentration of the reactants) is of about $10^9 - 10^{10}$ collisions per second, which should also be mutiplied by a steric factor $p = 10^{-4} - 10^{-5}$ (or smaller), corresponding to an usually negative (*cf.* section 1.2) entropy of activation, $\Delta S^{\neq}$. For two reacting molecules, fixed into a suitable mutual orientation in the reaction site of enzyme, $p = 1$ and the number of collisions undergone by a molecule corresponds to the frequency of some weak vibrations, $\nu \cong 10^{12} - 10^{13}$ s^{-1}. This effect of effective concentration and suitable orientation of the reactant in reaction site may account for an increase in reaction rate by a factor of 10^8 per enzyme molecule (equation 4.5d, for $e = a = b = 1$ mol l^{-1}). Another effect is the lowering of activation energy by forming some intermolecular bonds, which facilitate the reaction. Let us, for example, discuss the mutarotation of α-D, 2, 3, 4, 6-tetramethylglucose, catalysed by pyridone-2; the latter and the glucose ester appear, intermediary, as tautomeric forms (Figure 4.6). Other ways in which the enzyme can lower the activation energy are discussed by Kosower [33], and the importance of the simultaneous intervention of such intermolecular bonds, especially H-bonds (facilitated proton transfer), in a review by Wang [36]. The maximum value of rate constant for enzymic reactions corresponds to the reaction controlled by diffusion of reactants towards the reaction centre of the enzyme:

$$k_E = \frac{2N_A}{1000} \sigma_{BE} (D_B + D_E), \tag{4.6}$$

where σ_{BE} is the effective collision diameter of the enzyme-substrate reaction (approximately the size of the substrate molecule); D_B and D_E are diffusion constants of substrate and enzyme, respectively. If the enzyme and reactant bear charges of opposite sign, the limiting velocity k_E may even exceed the value given by equation (4.6) [34]. With $\sigma_{BE} = 2$Å and

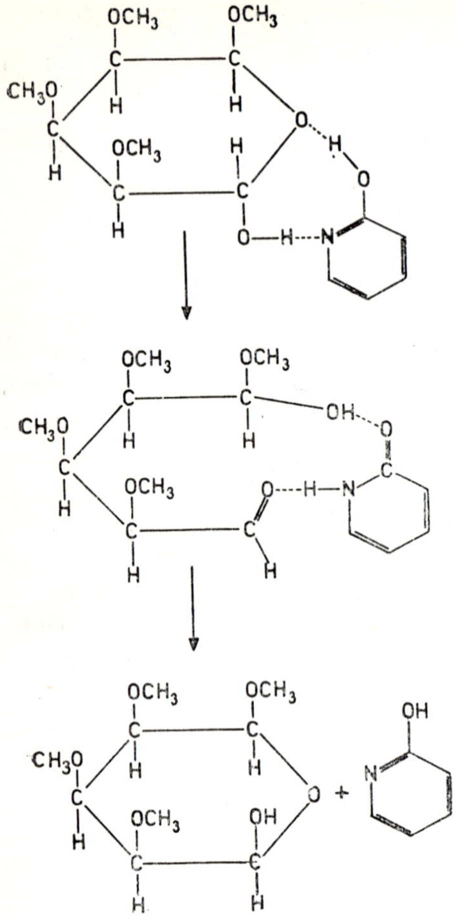

Figure 4.6. The catalysis of tetramethylglucose mutarotation by pyridine.

$D_B = 10^{-6}$ cm^2 s^{-1}, usual for micromolecules, the maximum values calculated according to (4.6) are of the order of $k_E = 2 \times 10^8$ s^{-1}.

A few examples: at 0 °C, 1 mole of Fe^{2+} or Fe^{3+} ions can decompose 10^{-5} moles of H_2O_2 per second; 1 mole catalase — 10^5 moles of H_2O_2 per second. The activation energy for the decomposition of H_2O_2 (into H_2O + + $^1/_2$ O_2), without catalyst, is of 18.0 kcal mol^{-1}, while in the presence of I^- ions, it is *ca.* 13.5 kcal mol^{-1}; with catalase E = 6.5 kcal mol^{-1}. For the hydrolysis of sucrose and urea, in acidic catalysis, the activation energies are *ca.* 25 kcal mol^{-1}, while in the presence of saccharose and urease they are of 11.0 and 12.5 kcal mol^{-1} respectively [37].

The study of enzymic reactions is basically very laborious, since the enzymic activity is strongly influenced by pH, the presence of various ions, thermal inactivation — if experiments are conducted at temperatures where denaturation of the enzyme begins, etc. [34].

The synthesis of an enzyme like synthetic polymer was also reported [67]. It is a polyethyleneimine with a molecular weight of 60,000 dalton with 10% of the NH residues alkylated by dodecyl groups ($C_{12}H_{25}$), which act as binding sites, and 15% of them by methyleneimidazoles, which act as a catalytic centre. The polymer "synzyme" accelerates 10^{12} times the hydrolysis rate of phenol sulphate esters, compared to non-bonded imidazole, and even 100 times compared to the enzyme arylsulphatase type IIA.

4.3.2. *Electronic aspects of enzymatic activity*

Electronic aspects are discussed extensively in the book of Pullman and Pullman [1b] on the basis of HMO calculations for co-ferments and several series of substrates. We will only discuss some examples of HMO calculations in relation to more advanced calculations.

As a first example we shall discuss enzymic hydrolysis. The hydrolases do not contain co-ferments. Their active centres are made up of some amino acid residues of the enzyme. HMO computations indicate a correlation between the positive charge of the hydrolysed bond in the substrate and the hydrolysis rate ([1], Chap. 17). Table 4.2 lists the relative hydrolysis rates of N-glycosidic bonds [38], correlated with the π-electronic charge at the N-atom, as calculated by the HMO method [1] and with the total charge at this atom, as calculated by the AVE—CI—SCF method [15]. The correlation is rather good, for the ρ_N^{π} figures yielded by HMO calculations, but completely absent for the ρ_N^{tot} figures of the advanced calculations. Possibly, the more polarisable π-electronic system is more directly implicated in this reaction, which would explain the absence of correlation of the rate *versus* total charge at the N atom. According to Pullman [1b] the easily hydrolysable bonds of biological interest have a bipositive character. Table 4.3 gives the electronic π charges, ρ_A^{π}, ρ_B^{π}, calculated for some easily hydrolysable bonds by the HMO method ([1], Chap. 18), as well as the total charges of the two atoms of these bonds, as calculated by more advanced methods. As in the case of the hydrolysis of the glucidic bond in nucleosides, the more advanced methods give

TABLE 4.2 **Hydrolysis of ribosides; correlation with MO parameters**

Substrate	*Relative rate*	ρ_N^π	ρ_N^{total}	MSD
adenosine	103	+0.407	—0.163	0
inosine	100	+0.419	—	0
guanosine	90	+0.406	—0.153	1
2,6-diaminopurine	76	+0.399	—0.170	1
cytidine	54	+0.361	—0.239	2
purineribosid	47	+0.408	—0.161	1
uridine	7	+0.311	—0.236	2
thymidine	4	+0.307	—0.193	3
uric acid riboside	1	+0.277	—0.208	2
xantosine	1	+0.414	—0.118	1
Linear correlation coefficient, *r*		0.67	0.02	0.57

ρ_N^π — π electron charge; HMO calculations

ρ_N^{total} — total electronic charge, AVE—CI—SCF

MSD — minimal steric difference as compared to adenosine, see section 6.2.1.

TABLE 4.3 **Easily hydrolysable bonds**

Substance	*Bond*	ρ_A^π ; ρ_B^π (HMO)	ρ_A^{tot} ; ρ_B^{tot} *(other methods)*
Carbonic acid ester	C—O	+0.362; +0.214	
Phenil phosphate	O—P	+0.209; +0.358	
Adenosine triphosphoric acid	O—P	+0.204; +0.364	
Acetylcholine	C—O	+0.272; +0.101	+0.354; —0.351 (AVE—CI—SCF [35])
			+1.41; —0.74 (Hoffmann MO [39])
			+0.39; —0.23 (CNDO/2 [39])
Peptide bond	C—N	+0.256; +0.141	+0.411; —0.311 (AVE—CI—SCF [35])
Barbituric acid	C—N	+0.252; +0.230	+0.416; —0.277 (AVE—CI—SCF [35])
Allantoine	C—N	+0.233; +0.121	+0.340; —0.277 (AVE—CI—SCF [35])

negative total charges for the N and O atoms, although the C atoms remain with rather high total positive charges.

Potential surface calculations by the CNDO/2 method including all valence electrons were performed for a model enzymic hydrolytic reaction: the acylation-desacylation of CH_3COOCH_3 and $CH_3CONHCH_3$ in the catalytic centre of α-chymotripsine [68]. This active centre consists of the His-57, Asp-102 and Ser-195 side chains, in fixed positions, as determined by x-ray difraction analysis. The potential curves calculated for hydrolysis agree with the experimental kinetics of hydrolysis and the calculated heights of energy barriers give a satisfactory explanation for the experimental activation energy. According to these computations, the decrease of activation energy in the enzymic reaction (as compared to the uncatalysed one) is largely due to variations in delocalisation energy and Koshland's "orbital steering" effect (*cf.* section 4.3.3) is also important: a deviation of only 6° in the O=C⟨ ... O angle (carbonyl in methyl acetate ··· O atom in the Ser-195 residue) rises the activation energy with 7 kcal mol^{-1}.

HMO calculations for the highest occupied and lowest empty molecular orbitals in co-enzymes of the respiratory chain indicate that the reduced forms of co-enzymes are always good electron donors ($\varepsilon_{HO} > \alpha + 0.5\,\beta$) while the oxidised forms are good acceptors ($\varepsilon_{LE} < \alpha - 0.5\,\beta$). Also, progressing along the respiratory chain from the substrate towards oxygen, ε_{HO} increases for the reduced forms, while ε_{LE} decreases for the oxidised forms, corresponding to positivation of redox potentials E_0 and to the tendency of electrons to flow from the substrate towards oxygen (O_2). Table 4.4 (according to Szent-Györgyi *et al.* [40]) illustrates this. The last enzymes in this chain are cytochromes, heminic complexes with Fe^{3+}/Fe^{2+} ion as electron transferring unit. The high spin Fe^{2+} complexes (four 3*d* electrons with parallel spins) are more easily oxidised by the paramagnetic oxygen molecule than low-spin Fe^{2+} complexes (with two or zero 3*d* electrons of parallel spins); the latter are more easily oxidised by electron transfer. Magnetic moment determinations demonstrate that cytochrome *a* and cytochrome *c* are low-spin complexes suitable to electron transfer. The reduced form of cytochrome a_3, the last before oxygen is a high-spin complex, but the oxidised form of cytochrome a_3 is an equili-

TABLE 4.4 **Redox properties and HMO parameters for coenzymes of the electro transporting chain**

Coenzyme	ε_{HO}	ε_{LE}	E_o, V
Activated substrate (SH_2/S)	—	—	—0.45
Nicotin adenine nucleotide			
NAD^+ (oxidised)	$\alpha + 1.032\beta$	$\alpha — 0.356\beta$	—0.32
NADH (reduced)	$\alpha + 0.292\beta$	$\alpha — 0.915\beta$	
Flavine mononucleotide			
FMN (oxidised)	$\alpha + 0.496\beta$	$\alpha — 0.343\beta$	—0.12
$FMNH_2$ (reduced)	$\alpha — 0.105\beta$	$\alpha — 0.949\beta$	
Cytochrome Fe^{3+} (oxidised)	—	$\alpha — 0.080\beta$	
Cytochrome Fe^{2+} (reduced)	$\alpha — 0.090\beta$	—	+0.26
O_2/O_2^{2-}	—	—	+0.82

brium mixture of the low- and high-spin forms and thus easily reduced also by electron transfer [41].

Another molecular orbital study, by the PPP-method for π-electrons (*cf.* section 2.6) deals with adenine and various bases and correlates the relative rates of ATP phosphorylation of the ribosides (reaction catalysed by adenosine kinase) with π-electron parameters for the respective heterocyles [69]. The relative rates, V_{rel} can be correlated to the frontier electrophilic electron density, f_3^e, (*cf.* section 2.3.4) at the N_3 atom of the base by the relation:

$$\log V_{rel} = —23.9 f_3^e + 2.1; \quad r = 0.9; \quad N = 13 \tag{4.6a}$$

This correlation, together with the existing enzymological data, suggest that adenosine gives a complex chelate with the co-factor Mg^{2+} in the *syn* configuration. Table 4.5 lists relative rates V_{rel}, Michaelis constants, K_M, the f_3^e-parameter and the minimal steric difference of nitrogen bases, as compared to adenosine (MSDs, *cf.* section 6.2.4). One should remark that f_3^e-parameter does not correlate with Michaelis constants, K_M, (the calculated π-electron charge on N_3 do not either [69]) which instead correlates to some extent with the general difference of shape against adenosine (MSD - figures, $r = 0.63$). If substitutions in riboside moiety are also considered (date in [69]) no correlation with either f_3^e or MSD is possible.

TABLE 4.5 **Electronic and rate parameters for adenosine kinase catalysed phosphorilations of some ribosides of purine bases (data listed by Kaneti [69])**

Nucleoside	V_{rel}	$K_M \times$ $\times 10^{-6}$(M)	f_3^e	*MSD*
9H-adenosine	1.00	1.6	0.0905	0
purinoside	3.45	78	0.0586	1
6Cl-purinoside	1.77	55	0.0825	0.5
$6CH_3$-purinoside	3.16	—	0.0640	0
$6CH_3O$-purinoside	2.09	—	0.0842	1
2F-adenosine	0.21	43	0.0990	1
2,6-diaminopurinoside	0.04	42	0.1028	1
$6CH_3S$-purinoside	2.44	—	0.0709	1.5
3H-adenosine	0.32	8000	0.1014	4
8-aza-adenosine	1.89	—	0.1045	0
$4NH_2$-formycin	3.11	—	0.1075	3
toyocamycin	1.05	8.0	0.0679	3
tubericidin	2.70	40	0.0760	2
N_1 oxide-adenosine	4.03	41	—	1
formycin	1.01	220	—	4
$6CH_3NH$-purinoside	1.50	1.6	—	1
$4CH_3S$-adenosine	0.34	140	—	1.50

As a general remark concerning electronic substrate parameters, it seems that if π-electronic parameters correlate somewhat with enzymic rates and other reaction characteristics, total charges calculated by all valence-electron methods do not seem to correlate at all. Where π-electronic and σ-electronic systems co-exist, the former seem to be more important for the reactions considered in this paragraph.

4.3.3. *Unusual explanations of enzymatic activity*

In the previous paragraph we have reviewed attempts to explain the high enzymic activity by means of mechanism similar to those encountered in homogeneous and heterogeneous non-enzymic catalysis; these mechanisms already belong with classical chemistry. These mechanisms operate through hydrogen bonding, electrostatic effects of some cations, which favour shifting of valence electrons towards the advancement of reaction or formation of rather unstable intermediary complexes, with atoms of co-enzymes of particularly high reactivity indexes for the corresponding reactions. Some authors, however, consider that these explanations are insufficient as to account for the peculiarly strong speeding up of some reactions under enzymic influence. We shall detail two unusual mechanisms in order to explain this acceleration: the orbital steering and "conformon".

Storm and Koshland's [42] orbital steering hypothesis affirm that the catalytic efficiency of enzymes (in increasing reaction rate up to 10^{20} times) depends not only on their capacity to bring together reacting atoms but on their ability of aligning orbitals along those directions which especially favour the reaction. These authors base their hypothesis on studies of esterification and γ-lactone-forming rates. The introducing of some cyclic systems into molecules entails the rigidisation of reacting group orientations and the fixation of certain mutual orientations of hydroxilic O atoms which onset carboxylic C.

In Figure 4.7 are shown the compounds whose lactone-forming ability was studied, as well as the reactions rates at 25 °C in ethanol solution. Even if the relative velocities were corrected for effects of group approaching, of twisting and of configurational isomerism, there would remain an unexplained speeding up of the value of 2×10^4. However the orbital steering effect should be sensed through an important change of overlap integrals for the orbitals of reacting atoms *versus* orientations. Such calculations on γ-hydroxyacids investigated by Koshland, have shown small variations of these overlap integral-increases, of only 10—20% for a rate increase 10^3 [43]. As a result of such calculations, it seems that the accelerating effect of enzymes should, anyway, be looked for in various "local concentration" effects, in the fixing of reactants in favourable orientations, in electrostatic charge effects, solvation and other "classical" effects.

Volkenstein's conformon hypothesis [45] takes into account the interaction between the fast motion of electrons and the slow configuration motion of nuclei - i.e., the polarisation of the medium throughout chemical reaction, like in Marcus' [46] theory for simple redox reactions (between complex ions of transition metals). This way the usual quantum

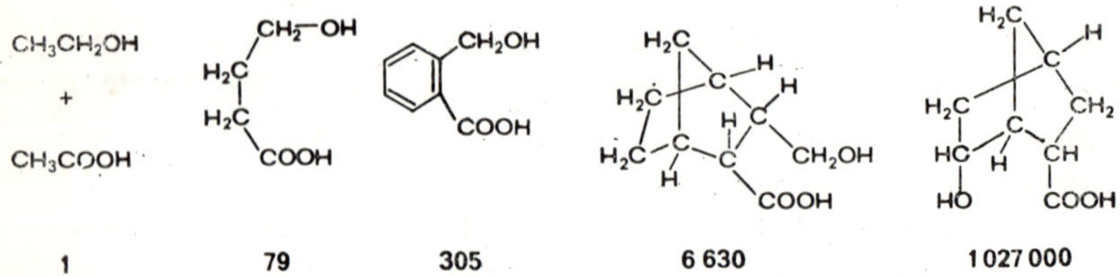

Figure 4.7. Relative reaction rates for lactone formation by γ-hydroxyacids.

chemical approximation, that is considering the motion of electrons separately from that of nuclei, is given up. Within this usual approximation one constructs a potential hypersurface for various intermediary (in reaction) nuclear configurations, by calculating the electronic energy of each nuclear configuration, for the nuclei in fixed positions, at rest. The interaction between electron and nucleo-configurational motion in biopolymers would consist of the following. The electronic transition within the reaction brings about a configurational change of macromolecules due to changes occurring on the free energy hypersurface (as a function of configuration, of nuclei positions). Due to the coupling of the two types of motion (in some ways resembling the polaron of solid state theory, whence the term conformon), the system follows a pathway of lower potential barrier than in absence of this coupling. According to Volkenstein the conformon concept may be most directly applied to enzymic redox reactions, like those catalyzed by cytochromes.

In addition, we can mention the theory advanced by Comoroşan [46a] for the mechanism and specificity of enzymic reactions, based on elements of quantum mechanical measurement theory. According to this theory, the enzyme is a measuring system, able to reveal far more delicate modifications of the substrate molecule, than usual physical methods.

4.4. Various biologically active compounds

4.4.1. *Carcinogenic agents.*

There is still doubt as to the change that makes a cell escape organismic control and divide at random. However, many harmful agents causing cancer are known: viruses, certain polycyclic hydrocarbons, some azo-derivatives, steroid hormone analogues, mutagenic agents, irradiation, hormonal overstressing, and even a simple metallic plate (or other inert material) implant. The sensitivity of the various tissues and of various animals to one and the same carcinogenic agent often differs within the extreme limits. Except for viruses, there also exists the requirement that the agent repeats its action throughout some time duration; however the malignant state occurs after a considerable time compared to the cell cycle. The acting mechanism of these agents is not known either but, evidently, they have to interact someway with a cell component.

There are scores of hypotheses on this mechanism. A good correlation between the mutagenic and the carcinogenic activity can be observed using x-rays, U.V.-radiations, urethans, N-yperites, while the strongly carcinogenic polycyclic hydrocarbons and azo-derivatives are not at all mutagenic ([47] Chap. 9, section 1). Nevertheless, the carcinogenic hydrocarbons metabolise into epoxides with both carcinogenic and mutagenic activities; a good parallel also exists between the epoxidic mutagenicity and the carcinogenic activity of hydrocarbons [48]. Other theories assume that carcinogenic agents only activate some pre-existing oncogenic viruses, in the cell [49]. Finally, several authors have found a parallel between the carcinogenic strength of the aromatic hydrocarbons or azo-dyes and their ability to react with certain protein classes; the action of alkylating mutagenic agents depends on their interaction with DNA ([50] Chap. 12). For this reason, at least within certain classes of agents, a correlation between the reactivity of certain molecular regions—i.e. between certain reactivity parameters calculated by the MO-method, and their carcinogenic activity is expected.

The most widely known chemical theory of carcinogenesis is the theory of *K*-regions of the polycyclic aromatic hydrocarbons, elaborated by the **Pullmans** [51]. The attaching of these hydrocarbons to their centres of action should be mediated by the most reactive region of the carcinogenic hydrocarbon. This is the *K*-region, defined by the C—C bond with the highest π-bond order and the largest free valence indices for the terminal atoms of this bond. Also defined is an *L*-region, consisting of the pair of C-atoms in *para*-position, with the highest affinity for addition reactions. In Figure 4.8 are shown the *K*- and *L*-regions for 1,2 benzanthracene. For the reactivity of the *K*-zone two indices are used: the first one—sum of the π-bond index and free valence indices of the two valence terminal C-atoms, conventionally designated by R_K; the second index applicable to both regions, is the localisation energy of the π-electrons to the atoms involved in the bond plus the localisation energy of the π-bond to the corresponding atoms; it will be labelled by EL_K and EL_C, respectively. The *in vitro* chemical study of these hydrocarbons reveals this *L*-region as generally being more reactive than the *K*-

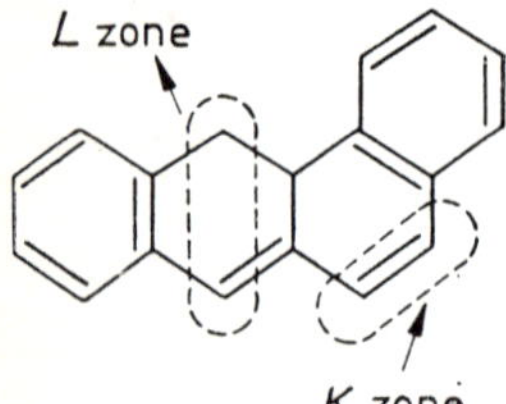

Figure **4.8.** The *K*- and *L*-regions for benzanthracene.

region, for most of reactions involving these hydrocarbons. The *K*-region takes part in some more specific reaction, e.g. the fixation of OsO_4. If this would also be the case *in vivo*, we could expect the requirement that *L*-region is not too reactive to arise in addition to that of *K* being sufficiently reactive; this allows *K*-region to act prior to degrading of molecules by other mechanisms which involve the *L*-region.

The *K*-region rule for polycyclic hydrocarbons provides for R_K by being larger than 1.28 or EL_K by being smaller than 3.31, and at the same time, EL_C by being not too small. This explains, e.g. why 8-methyl-3,4-benzophenanthrene—although its R_K is 1.309—has a smaller carcinogenic activity than 1,2-benzanthracene with $R_K = 1.283$. If one also considers the not too-highly reactivity requirement for the *L*-zone, R_K and EL_K are quite well correlated with carcinogenic activity, as far as this activity can be defined quantitatively [50, 51]. More difficult to explain is the carcinogenic activity of some derivatives substituted within the *K*-region of these hydrocarbons, e.g. the rather high carcinogenicity of 4-fluoro-10-methyl-1,2-benzanthracene; the substituents for *K*-region should render far more difficult the reactions of fixation to these regions [52]. The R_K, EL_K etc. parameters have been calculated by means of HMO method [51]. For azo-dyes and derivatives of stilbene with carcinogenic properties (for example, dimethylbenzanthracene - DAB, also called butter-yellow; *cf.* Figure 4.9), the *K*-region would consist of the —N=N- or —CH=CH-groups bonding together the two aromatic rings. For the —N=N-group the fixation ability depends on the nucleophilic reactivity of the two N-atoms, which will parallel the energy increase (*EP*) of the π-electronic system in the protonation of the two N-atoms (the transfer from $\alpha_N = \alpha + 0.5\,\beta$ to $\alpha_{N+} = \alpha + 1.5\,\beta$). The correlation between this *EP* index and carginogenicity is rather bad; a parallelism between *EP* and carcinogenic activity is only found for the 5 methyl-substituted derivatives on the benzene rings [53].

Allison and co-workers [54] have advanced the theory that the carcinogenic properties depend upon the electron-donor and acceptor capa-

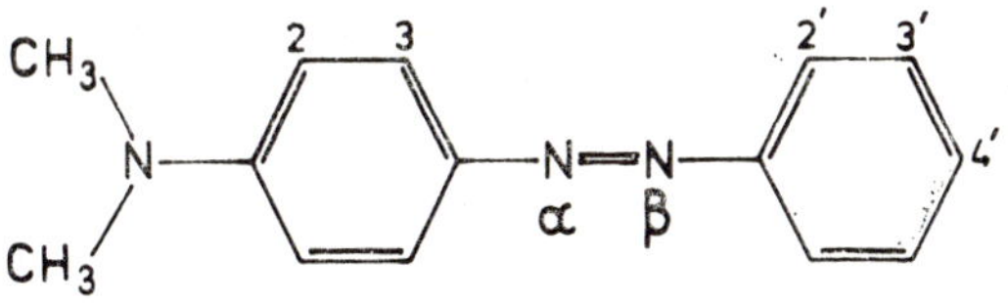

Figure **4.9.** Numbering of atoms in DAB (dimethylaminoazobenzene).

city of the compound. This theory is parallelled by a metabolic one — the carcinogenic agents intervene into the electron-transfer chain from the oxidative phosphorylation process, inhibiting it therefore; the cells which acquire the highest energy fraction through glycolysis would be favoured. During certain natural processes of proliferation, those cells would be selected whose metabolic characteristics (aerobic glycolysis) are specific to malignant cells [55]. The easiness of electron donation and acception — characterised by the least positive oxidation polarographic potential E_1^{ox}, and the least negative reduction polarographic potential E_1^{red} — is well correlated to carcinogenic activity. Measurements of the polarographic steps E_1^{ox} and E_1^{red} were carried out in solutions of acetonitrile; evidently, comparisons may be done only within certain classes of similar compounds, whose solvation energy variations (on oxidation and reduction) are the same for all the compounds of the class of interest. Table 4.6 gives a comparison of some reactivity parameters: R_K, EL_K for hydrocarbons, EP for azo-derivatives, the polarographic oxi-

TABLE 4.6 **The carcinogenic activity and electronic parameters of some carcinogenic hydrocarbons and azo-derivatives**

Compound	*Carcino-genic activity*	*) R_K / EP	EL_K (β units)	EL_L	E_1^{ox} (V)	E_1^{red}
Benzene	—	—	4.07	6.54	2.38	unred.
Naphthalene	—	—	3.56	5.98	—	—
Anthracene	—	—	3.53	5.38	1.19	—2.07
1,2-benzanthracene	++	1.283	3.29	5.53	1.33	—2.11
3,4-benzacridine	—	1.260	—	—		
Azobenzene	—	—	—	—	1.89	—1.45
p-dimethylamino-azobenzene	++	—0.670 *	—	—	0.86	—1.58
2-methyl-*p*-dimethyl-aminoazobenzene	—	—0.675 *	—	—	0.80	—1.60
3′-methyl-*p*-dimethyl-aminoazobenzene	+++	—0.748 *	—	—	—	—
3′-methoxy-*p*-dimethyl-aminoazobenzene	+++	—0.670 *	—	—	—	—
3′-F-*p*-dimethylamino-azobenzene	+++	—0.670 *	—	—	0.90	—1.50
1,2-benzpyrene	++++	—	3.23	—	1.10	—1.95

*) Column three lists reactivity parameters, R_K for the *K*-region, *EP* (values with asterisk) for the analogous region in azodyes. For significance of K_R and *EP* — see section 4.4.1.

dation and reduction steps E_1^{ox} and E_1^{red}, as well as the carcinogenic activity semiquantitatively characterised by one or more "+"; one or more "—" do the same with non-carcinogenic compounds.

As the intimate mechanism of carcinogenesis is not known, little is also known about structural and reactivity characteristics which determine the carcinogenic activity of a compound. Certain conclusions pertaining to the physico-chemical mechanism of carcinogenesis, may, nevertheless, be drawn [56]. In order that a polycyclic aromatic hydrocarbon be carcinogenic, it has to present a good electron transfer (acceptor and donor) capacity, a geometry as close as possible to that of nucleic acids base-pairs and a molecular thickness not exceeding 4 Å. Also, a large number of various types of carcinogens may be considered as electrophilic reactants—either themselves or some of their metabolic products. Thus, the cell "target" involved in carcinogenesis would show preference to such nucleophilic centres.

4.4.2. *The energy-rich compounds*

Energy-rich compounds usually are biological compounds which have for hydrolysis reaction an intensely negative ΔG. Owing to this fact, their indirect hydrolysis—along a sequence of processes—may concommitantly bring about the advancing of some reactions with positive ΔG. Schematically with adenosine triphosphate as energy-rich compound:

$$\mathrm{ATP} + \mathrm{H_2O} \rightarrow \mathrm{ADP} + \mathrm{H_2PO_4^-}; \quad \Delta G' \cong -7.5 \text{ kcal mol}^{-1} \text{ at pH} = 7 \tag{4.7}$$

and an endergonic reaction

$$\mathrm{B} + \mathrm{C} \rightarrow \mathrm{D} + \mathrm{H_2O}; \quad \Delta G'' > 0, \quad \text{yet} \quad \Delta G' < -\Delta G'' \tag{4.7a}$$

the two reactions may be coupled, by means of an enzyme E, according to a mechanism of the type:

$$\begin{array}{c} \mathrm{ATP} + \mathrm{B} + \mathrm{E} \rightarrow \text{complex} \\ \underline{\text{complex} + \mathrm{C} \rightarrow \mathrm{E} + \mathrm{D} + \mathrm{ADP} + \mathrm{H_2PO_4^-}} \\ \mathrm{ATP} + \mathrm{B} + \mathrm{C} \rightarrow \mathrm{D} + \mathrm{ADP} + \mathrm{H_2PO_4^-} \\ \Delta G_{\text{total}} = \Delta G' + \Delta G'' < 0 \end{array} \tag{4.7b}$$

The energy-rich compounds are generally anhydrides of the phosphoric acid with another acid. Qualitatively, in the light of direct valence bond

(VB) theory, the exergonic character of hydrolysis of energy-rich compounds may be explained by the existence in the hydrolysis products of several equi-energetic boundary structures which leads to a conjugation energy higher than that of reactants. For example—in the hydrolysis of acetylphosphate (the ionic formulas at pH = 7 are given):

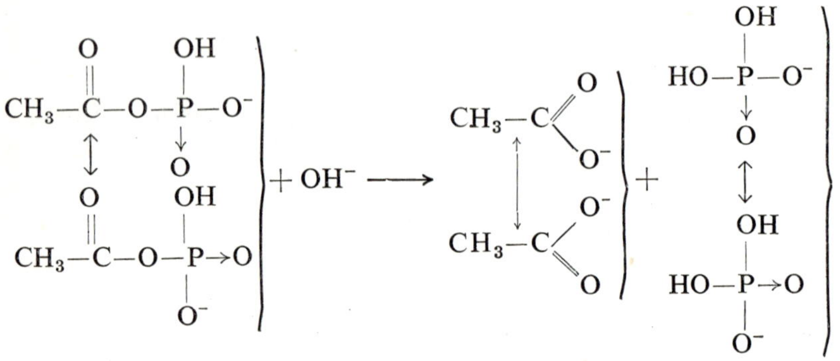

$$\left\{\begin{array}{c} CH_3-\overset{\overset{O}{\|}}{C}-O-\overset{\overset{OH}{|}}{P}-O^- \\ \downarrow \\ O \\ \updownarrow \\ CH_3-\overset{\overset{O}{\|}}{C}-O-\overset{\overset{OH}{|}}{P}\rightarrow O \\ | \\ O^- \end{array}\right\} + OH^- \longrightarrow \left\{\begin{array}{c} CH_3-C\begin{matrix} \nearrow O \\ \searrow O^- \end{matrix} \\ \updownarrow \\ CH_3-C\begin{matrix} \nearrow O^- \\ \searrow O \end{matrix} \end{array}\right\} + \left\{\begin{array}{c} \overset{OH}{|} \\ HO-P-O^- \\ \downarrow \\ O \\ \updownarrow \\ \overset{OH}{|} \\ HO-P\rightarrow O \\ | \\ O^- \end{array}\right\} \tag{4.8}$$

Acetylphosphate has two equi-energetic boundary structures, while each of the acetate and phosphate ions have two, i.e. in all 4 boundary structures. In the case of methyl-phosphate hydrolysis (an "energy-poor" compound):

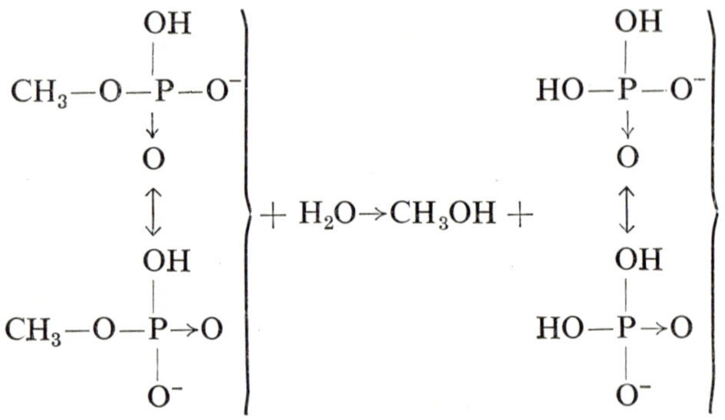

$$\left\{\begin{array}{c} \overset{OH}{|} \\ CH_3-O-P-O^- \\ \downarrow \\ O \\ \updownarrow \\ \overset{OH}{|} \\ CH_3-O-P\rightarrow O \\ | \\ O^- \end{array}\right\} + H_2O \rightarrow CH_3OH + \left\{\begin{array}{c} \overset{OH}{|} \\ HO-P-O^- \\ \downarrow \\ O \\ \updownarrow \\ \overset{OH}{|} \\ HO-P\rightarrow O \\ | \\ O^- \end{array}\right\} \tag{4.8a}$$

both the methyphosphate and phosphate ion have two boundary structures each. These differences should also be mirrored in HMO calculations, the hydrolysis products being expected to have a localisation energy of π-electrons higher than the corresponding energy-rich compounds. In Table 4.7 (according to Pullmans' book [1] Chap. VII), are given the free enthalpies for the hydrolysis of some energy-rich phosphates and of an "energy-poor" phosphate, as well as the increases of delocalisation energies for π-electrons, ΔDE_π, on hydrolysis, in β units and in kcal mol^{-1}

TABLE 4.7 **The calculation of ΔG for hydrolysis of energy-rich phosphates**

Compound	$-\Delta G$ *experi-mental* (kcal mol^{-1})	ΔDE_π (β units; kcal mol^{-1})	$-\Delta RE$ (kcal mol^{-1})	*Other effects* (kcal mol^{-1})	$-\Delta G_{cal}$ (kcal mol^{-1})
$RCH_2OPO_3^{2-}$	3	—0.003; 0.05	0.0	—	2.95
$RCOOPO_3^{2-}$	10—12	—0.036; 0.59	0.7	3.2	6.0
COOH | $H_2O{=}C{-}OPO_3^{2-}$	11.5—12.5	—0.010; 0.16	0.5	9	11.6
NH ‖ $H_2N{-}C{-}NHP_3^{2-}$	9—10	—0.018; 0.30	0.7	7	9.5
$RCH_2OPO_2^{-}{-}OPO_3^{2-}$	7—8	—0.095; 1.57	1.4	—	5.9
$RCH_2OPO_2^{-}{-}OPO_2^{-}{-}OPO_3^{2-}$	7—8	—0.176; 2.90	2.0	—	8.7

(for the usual value, $\beta = -16.5$ kcal mol^{-1}). Thus, for the hydrolysis of ATP, $\Delta DE_\pi = -3$ kcal mol^{-1}, while $\Delta G = -7.5$ kcal mol^{-1} (at pH = 7) compared to $\Delta G = -3$ kcal mol^{-1} for hydrolysis of energy poor phosphates. The ΔDE_π variation of the delocalisation energy could not account for the entire value of ΔG in the hydrolysis of energy-rich phosphates, and especially, it does not explain the negative ΔG for the hydrolysis of the "poor" ones. From the sum of bond energies one should obtain $\Delta H \cong 0$. The hydrolysis does not modify the total number of particles in solution; the increase of mixing entropy ΔS_{mix} would be (*cf.* Appendix 1), of *ca.* +9.8 cal deg^{-1}. For $T = 300$ K, $-T\Delta S = -2.9$ kcal mol^{-1}, which would explain the value of ΔG for the hydrolysis of "energy-poor" phosphates.

For energy-rich phosphates a difference of *ca.* —1.5 kcal mol^{-1} (for ATP) remains to be explained. Examination of molecular diagrams of these compounds (calculated by HMO [1], Chap. 7) shows that some energy-rich compounds possess charged atoms, close to each other which may lead to an electrostatic repulsion, lowered by hydrolysis that takes apart the two groups (*cf.* Figure 4.10). These energy differences due to electrostatic repulsion [57], ΔRE (*cf.* [1] Chap. 7) have also been calculated and are given, too, in Table 4.7. Sometimes other effects may occur: in the hydrolysis of carboxyl phosphates a new ionisable group appears (the carboxyl whose ionisation, at pH = 7 corresponds to $\Delta G = -3.2$ kcal mol^{-1}) while for the hydrolysis of phosphoenolpyruvate by recovery of pyruvic acid to ketotautomeric (more stable) form, $\Delta G = -9$ kcal mol^{-1}.

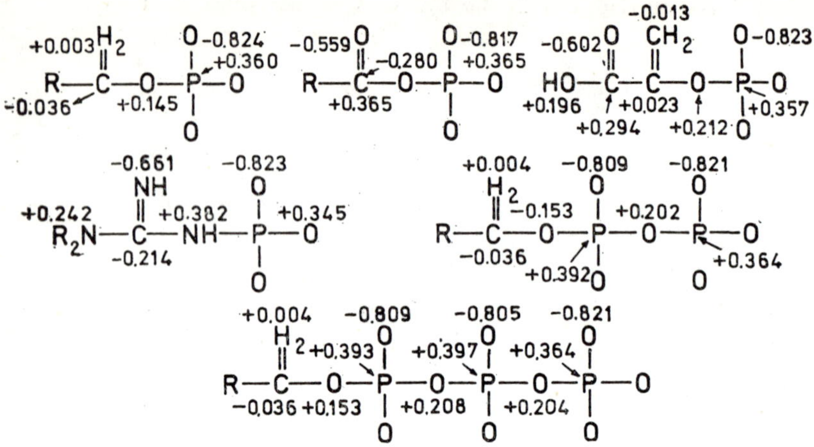

Figure **4.10** π-electron atomic charges of some energy-rich phosphates.

TABLE 4.8 **Electronic characteristics and bacteriostatic activity of sulphamide drugs**

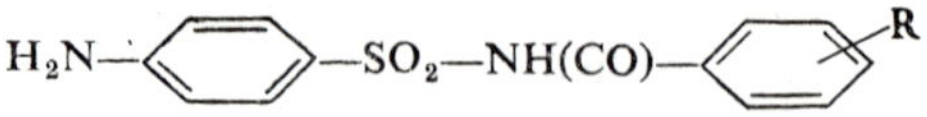

H_2N—⟨benzene⟩—SO_2—NH(CO)—⟨benzene⟩—R

Substituent	log (1/*Cr*)	c_{lN}	ρ_N	ρ_N^2
	Anilines (ρ_c = 0.00)			
4-NH_2	4.35	0.383	0.077	0.006
4-CH_3O	4.47	0.453	0.081	0.007
4-CH_3	4.57	0.453	0.081	0.007
H	4.80	0.477	0.083	0.007
4-Cl	4.80	0.459	0.082	0.007
4-NO_2	5.85	0.484	0.106	0.011
	Benzamides (ρ_c = 0.20)			
3-CH_3, 4-CH_3O	5.25	0.041	0.148	0.022
4-CH_3	5.40	0.035	0.148	0.022
3,4-CH_3	5.40	0.042	0.148	0.022
4-CH_3	5.40	0.036	0.148	0.022
3-CH_3	5.40	0.018	0.149	0.022
H	5.25	0.000	0.148	0.022
4-Cl	5.10	0.027	0.148	0.022
4-CN	4.05	0.000	0.151	0.023
4-NO_2	4.50	0.000	0.153	0.023

To the calculated ΔG was also added $-T\Delta S_{\text{mix}} = -2.9$ kcal mol^{-1}. The agreement is satisfactory, except for carboxylphosphate. The reason: in the resulting carboxyl ion, the two O atoms have $-\frac{1}{2}$ e charge that would correspond to a value of α much less electronegative than the value $\alpha_0 = \alpha + 2\beta$ employed in calculations for ΔDE_π; with much less electronegative α_0 values, i.e. closer to α_c, one obtains a delocalisation energy increased by *ca.* 3 kcal mol^{-1}.

Calculations of all valence electrons with the EHT method (*cf.* section 2.4) indicate that the property of the energy-rich and energy poor of biological phosphates correlates with the total charge of the estheric C-atom [70].

4.4.3 *The quantum perturbation theory of drug action*

This is reviewed by Cammarata [58]. This theory is based on the formation of a drug (S)-receptor (R) complex:

$$S + R \overset{a}{\rightleftarrows} SR \overset{b}{\rightarrow} \text{response (A)} \tag{4.9}$$

in which the determining step for the biological response is either the association-dissociation step (a), according to Paton' s kinetic theory or step (b), according to Ariens' occupational theory. In both cases the biological (pharmacological) response A will be determined by the drug-receptor interaction, i.e. by the variation ΔE^e of electron energy, ΔE^d of solvation energy, ΔE^s of steric energy and ΔE^p of conformational energy (*cf.* section 3.1.3):

$$A = -\frac{1}{RT}(\Delta E^e + \Delta E^d + \Delta E^s + \Delta E^p) + d \tag{4.10}$$

For the significance of the pharmacological response-biological activity, A, see also Appendix 3. The term d represents a constant. The terms ΔE^e and ΔE^d may be expressed by electronic structure parameters (*cf.* section 3.1.3) and usually one assumes that only these two terms are modified within a class of drugs. Structure-activity correlations within such a class are established mostly through parameters related to ΔE^e change.

As the expressions of ΔE^e and ΔE^d can be written as sums over interactions between pairs of atoms (on S and R, respectively), for a given receptor R, the pharmacological activity and the ΔE^e, ΔE^d terms may be additive with respect to groups of atoms on the drug molecules, S. This additivity seems rather well verified, as indicated by the studies of Belleau and Lacasse [58a] concerning the conformational modifications of the cholinergic receptor and by those of Cammarata and Yau [58b] concerning the antibacterial activity of a group of tetracyclines (Figure 4.11). Thus, on comparing acetylcholinesterase inhibitors I and II (Figure 4.11) the free enthalpy difference $\Delta(\Delta G)$ for the two drug receptor interactions should be given by dehydration of CH_3 group (—0.73 kcal mol^{-1}) and by the van der Waals interaction of two CH_3 groups (—0.60 kcal mol^{-1}); the assumption is made that inhibitors' interaction with the enzyme occurs through methyl-methyl interaction. Compound II is a methylated derivative of I. The $\Delta(\Delta G)$ difference between the free binding enthalpies of the two groups is—1.355 kcal mol^{-1}, very close to the theoretical figure of 1.33 kcal mol^{-1} [58a]. Similarly, for a series of nine tetracycline derivatives tested upon *Escherichia coli* and *Salmonella typhimurium* bacteria, the linear correlation coefficient between the experimental activities and those calculated by an optimized additive formula is *ca.* $r = 0.90$ [58b].

Equations (4.11a and 4.11b) of section 3.1.3 for ΔE^e and ΔE^d cannot be applied directly to pharmaceutic activity-electronic parameter correlations as they also contain parameters of the biological receptor, whose structure is not known. If the S-R interaction is controlled by fron-

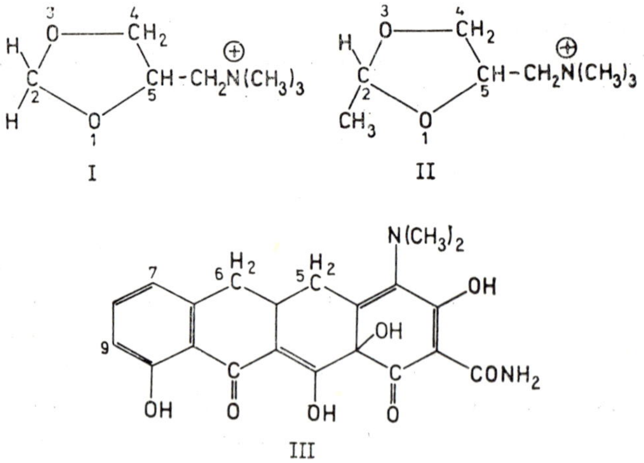

Figure **4.11**. — Molecular structure and numerotation of atoms for two acetylcholinesterase inhibitors (I and II) and for tetracycline skeleton (III).

tier orbitals, equation (4.11a) section 3.1.3 for ΔE^e becomes, as shown by Klopman and Hudson [61]:

$$\Delta E^e = \sum_{\alpha}\sum_{\beta'}(2c_{h'\beta'}\,c_{l\alpha} - 2c_{l'\beta'}c_{h\alpha})\,\beta \tag{4.11}$$

and if the receptor parameters $c_{h'\beta'}$ and $c_{l'\beta'}$ are considered as constants K_α and K'_α:

$$\Delta E^e = \sum_{\alpha}(K_\alpha c_{l\alpha} - K'_\alpha c_{h\alpha}) \tag{4.11a}$$

with $c_{l\alpha}$ and $c_{h\alpha}$ the coefficients of the atomic orbital α in the lowest empty and highest occupied molecular orbitals of the drug S. Likewise, one obtains for ΔE^d

$$\Delta E^d = \sum_{\alpha}(\lambda_\alpha \rho_\alpha \pm \lambda'_\alpha \rho^2_\alpha) + d' \tag{4.11b}$$

in which the parameters of the receptor are included in λ_α and λ'_α. Equation (4.11b) for ΔE^d is valid if the interaction of frontier orbitals implies a complete charge transfer. The final formula for correlating the activity A with parameters of electronic structure becomes:

$$A = \frac{-1}{RT}\sum_{\alpha}(K_\alpha c_{l\alpha} - K'_\alpha c_{h\alpha} + \lambda\rho_\alpha \pm \lambda'\rho^2_\alpha) + d'' \tag{4.12}$$

If the S—R′ interaction is esentially electrostatic, without charge transfer, equation (4.11b) section 3.1.3 indicates that there are no changes in solvation energy and equation (4.12) becomes (only ΔE^e changes are considered):

$$A = -\frac{1}{RT}\sum_{\alpha}(\nu_\alpha \rho_\alpha + \xi_\alpha S^N_\alpha + \zeta_\alpha S^E_\alpha) + d'' \tag{4.13}$$

in which ρ_α is the π-electron charge density of atom α, S^E_α and S^N_α the delocalisability indexes of Fukui:

$$S^E_\alpha = \sum_{p,\,\text{occupied}} \frac{2c^2_{p_\alpha}}{\varepsilon_p - \varepsilon_\alpha}; \quad S^N_\alpha = \sum_{q,\,\text{free}} \frac{2c^2_{q_\alpha}}{\varepsilon_q - \varepsilon_\alpha}, \tag{4.13a}$$

and ν_α, ξ_α, ζ_α the coefficients adjusted so as to maximise the r-coefficient (*cf.* Appendix 4).

The interaction of nicotinic acid derivatives (Figure 4.12) with acetylcholinesterase seems to be controlled by charge-transfer interactions between frontier orbitals. The inhibiting action of these derivatives corre-

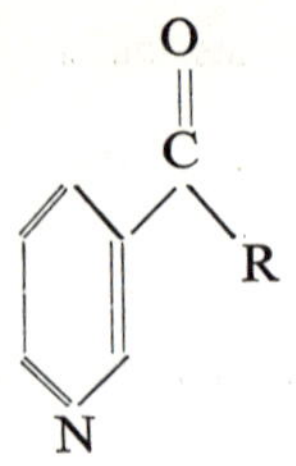

Figure **4.12**. Nicotinic acid derivatives.

R: OH, NH_2, CH_3, OC_2H_5

$p\,I_{50} = 5.32c_{lc}^2 - 1.24.$

lates to charge density c_{lc}^2 of the lowest empty orbital at the carbonylic carbon atom [62]:

$$A = pI_{50} = 5.32\ c_{lc}^2 - 1.24; \qquad r = 0.87,\ N = 4 \tag{4.14}$$

Here, pI_{50} is the logarithm of concentration, producing a 50% inhibition of the enzyme, taken with the opposite sign.

The same type of interaction seems to occur in two series of sulphonamide drugs. The minimal inhibiting concentration C_r upon *Escherichia coli* correlates with the coefficient c_{lN} of the iminic N atom in the lowest empty orbital and with the π-charge ρ_N of this atom:

$$A \equiv -\log C_r = 8.76\, c_{lN} + 319\rho_N - 1731\rho_N^2 + 45.4\rho_c - 13.28 \tag{4.14a}$$
$$r = 0.94,\ \ N = 15$$

The π-charge ρ_c of the carbonylic atom is taken to be equal to 0.20 for benzamide derivatives and zero for aniline derivatives. The high value of the coefficient of ρ_N^2 indicates a high influence of the solvation effects. The electronic parameters for this series were obtained by HMO calculations for the anilines and benzenamides from which the sulphamides derive [58a]. The correlation is illustrated by Table 4.8.

A cholinergic system controlled by electrostatic interactions are the trimethylphenylammonium derivatives of Figure 4.13. Total atomic

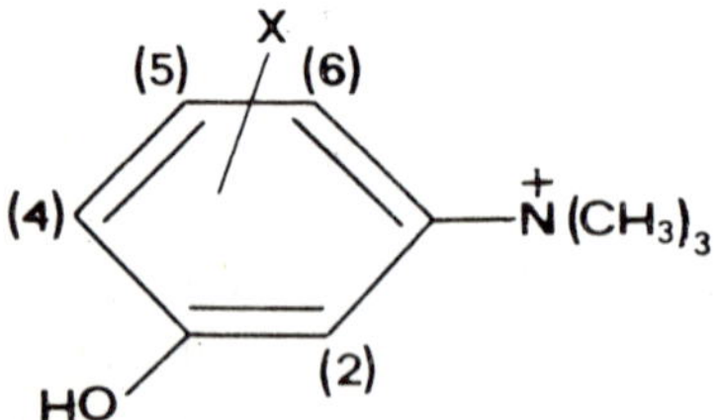

Figure **4.13**. Aromatic inhibitors of acetylcholinesterase.

charge calculations by the HMO and Del Re methods (*cf.* section 2.5) allowed the formation of the following correlation equation:

$$A \equiv pI_{50} = -257\, \rho_{O_3} - 283\, S^E_{O_3} - 64\, S^N_{O_3}; \qquad r = 0.98, \quad N = 6 \tag{4.14b}$$

The correlation may be explained by a hydrogen bond between the phenolic OH group and the cholinergic receptor [63].

The *in vitro* bacteriostatic activity of 18 tetracyclines (Figure 4.11, structure III) can be correlated with electronic parameters calculated by the HMO and Del Re methods, according to:

$$A = 56.17\, \rho_{O_{10}} + 16.19\, S^E_{O_{10}} + 48.79\, \rho_{O_{11}} - 1.10\, S^E_{O_{11}} + 71.32\, \rho_{O_{12}} + 18.36\, S^E_{O_{12}} + 3.38\, \rho_{O_6} + d; \quad r = 0.99,\ N = 18 \tag{4.14c}$$

This result agrees with a bacteriostatic mechanism in which the protein synthesis is inhibited by bonding to activated ribosomes, eventually *via* chelation of a metabolic cation [58c].

Several correlational equations were set up also by means of the σ-Hammet and the hydrophobicity π-constants. For a review, see [71]. As an example, the inhibition of malate dehydrogenase by substituted phenols (meta derivatives are not included) can be correlated by the equation [72]:

$$A = 0.73\, \sigma^2 - 0.65\, \sigma - 0.77\, \pi^2 + 2.44\, \pi + 3.47 \qquad r = 0.95 \tag{4.15}$$

As a conclusion, the work of Cammarata and his co-workers on the quantum theory of drug action, as well as Hansch and his co-workers (*cf.* section 3.3.7) on correlation of biologic activities with σ- and π-constants give high correlation coefficients ($r \geqslant 0.95$) but usually only for small series, only for a limited variation of chemical structure. The drug-receptor interaction is probably determined also by other factors which are not accounted for by electronic parameters and which do not change appreciably over a narrow range of a structural modifications. This is not surprising as the steric conformation of a molecule is very important for its biological activity (*cf.* section 6.2) and even the different types of intermolecular forces (*cf.* section 3.3) cannot always be related to electronic structure parameters.

REFERENCES

1. B. Pullman and A. Pullman, *Quantum Biochemistry*, Ch. 6; (1a) ch. 5; (1b) Part 3; (1c) B. Pullman, *Advan. Quantum Chem.*, **21**, 267 (1968).
2. A. Szent-Györgyi, *Nature*, **148**, 157 (1941); A. T. Vartanian, *Zh. Fiz. Khim.* (*J. Phys. Chem.*), **32**, 769 (1948); D. D. Eley, *Nature*, **162**, 819 (1948).
3. Iu. A. Vladimirov, *Photochemistry and Luminiscence of Proteins* (in Russian), Publ. House for Science, Moscow, (1965); A. D. Mc Laren and D. Shugar, *Photochemistry of Proteins and Nucleic Acids*, Oxford Univ. Press, (1964).
4. M. G. Evans and J. Gergelyi, *Biochim. Biophys. Acta*, **3**, 188 (1949).
5. M. Suard, G. Berthier and B. Pullman, *Biochim. Biophys. Acta*, **53**, 254 (1961).
6. M. H. Cardew and D. D. Elley, *Discuss. Faraday Soc.*, **27**, 113 (1959).
7. D. D. Eley and D. Spivey, *Nature*, **188**, 724 (1960).
8. A. Imamura, M. Kodama, V. Tagashira and Ch. Nagata, *J. Theoret. Biol.*, **10**, 356 (1966).
9. B. Maigret, B. Pullman and M. Dreyfuss, *J. Theoret. Biol.*, **26**, 321 (1970); B. Maigret, D. Perahia, and B. Pullman, *idem*, **29**, 275 (1970).
10. F. K. Winkler and J. D. Dunitz, *J. Molec. Biol.*, **59**, 169 (1971).
11. B. H. Zimm and N. R. Kallenbach, *Ann. Rev. Phys. Chem.*, **13**, 171 (1962).
12. D. D. Eley, in: *Horizons in Biochemistry*, B. Pullman and M. Kasha (eds.),- Academic Press, New York, (1962) p. 241.
13. P. Douzou, J. Francq and M. Hauss, *J. Chim. Phys.*, **58**, 926 (1961); R. O. Rahn, R. G. Shulman and J. W. Longworth, *J. Chem. Phys.*, **45**, 2955 (1966); *Proc. Nat. Acad. Sci. U.S.A.*, **53**, 893 (1965).
14. J. Ladik, *Quantum Biochemistry* (in Hungarian), Budapest, (1967).
15. S. Fraga and H. Carbo, *Technical Report*, TC-6914 (1969) Division of Theoretical Chemistry, Department of Chemistry, University of Alberta.
16. T. Nakajima and A. Pullman, *J. Chim. phys.*, **55**, 793 (1958).
17. J. P. Kokko, J. H. Goldstein and L. Mandell, *J. Amer. Chem. Soc.*, **83**, 2909 (1961).
18. H. T. Miles, *Proc. Nat. Acad. Sci. U.S.A.*, **47**, 791 (1961); R. V. Wolfenden, *J. Molec. Biol.*, **40**, 307 (1969).
19. H. Fujita, A. Imamura and Ch. Nagata, *J. Theoret. Biol.*, **28**, 143 (1970).
20. B. I. Sukhorukov, V. I. Poltev, V. R. Fedkina and Y. K. Knobel, *Dokl. Akad. Nauk SSSR (Proc. U.S.S.R. Acad. Sci.)*, **175**, 948 (1967).
21. R. V. Wolfenden, *J. Molec. Biol.*, **40**, 307 (1969); L. Gatlin and J. Davis, *J. Amer. Chem. Soc.*, **84**, 4464 (1967); A. R. Katrizky and A. J. Waring, *J. Chem. Soc.*, 1540, (1962); 3046, (1963).
22. E. Freese, *J. Theoret. Biol.*, **3**, 82 (1962).
23. P. O. Lowdin, in: *Electronic Aspects in Biochemistry*, B. Pullman (editor), Academic Press, New York, (1964), p. 167; *Rev. Mod. Phys.*, **35**, 724 (1963).
24. P. O. Lowdin, *Advan. Quantum. Chem.*, **2**, 213 (1967).
25. R. Rein and J. Ladik, *J. Chem. Phys.*, **40**, 2466 (1964).
26. J. Ladik, *Preprint QB8*, Uppsala Quantum Chemistry Group (1963).
27. R. Rein and F. Harris, *J. Chem. Phys.*, **41**, 3393 (1964); **43**, 4415 (1965); S. Lunell and G. Sperber, *Preprint QB32*, Uppsala Quantum Chemistry Group (1966).

28. J. Duchesne, J. Dépireux, A. Bertinchamps, N. Cornet and J. M. Van der Kaa, *Nature*, **188**, 405 (1960).
29. T. A. Hoffman and J. Ladik, *Advan. Chem. Phys.*, **7**, 8 (1964).
30. J. Ladik, D. K. Rai and K. Appel, *Preprint QB31*, Uppsala Quantum Chemistry Group (1966).
31. B. F. Rozsnay, F. Martino and J. Ladik, *J. Chem. Phys.*, **52**, 5708, (1970).
32. N. Sklenar, J. Ladik and G. Biczo, *Studia Biophys.*, **2**, 459 (1967); B. F. Rozsnai and J. Ladik, *J. Chem. Phys.*, **52**, 5711; **53**, 4326 (1970).
33. E. M. Kosower, *Molecular Biochemistry*, Part III, McGraw-Hill Book Co., New York, (1962).
34. J. L. Webb, *Enzymes and Metabolic Inhibitors*, Academic Press, London, (1963) Ch. 2.
35. S. Fraga and E. Fraga, University of Alberta, *Division of Theoretical Chemistry*, Tehnical Report TC-7001 (1970); 35a TC-7102 (1971).
36. J. H. Wang, *Science*, **161**, 328 (1968).
37. H. G. Bray and K. White, *Kinetics and Thermodynamics in Biochemistry*, J. & A. Churchill Ltd., London, (1957) Ch. VII.
38. Y. Takagi and B. L. Horecker, *J. Biol. Chem.*, **225**, 77 (1959).
39. D. Ajo, M. Bossa, A. Damiani, R. Fidenzi, S. Gigli, L. Lanzi and L. Lapiccirella, *J. Theoret. Biol.*, **34**, 15 (1972).
40. A. Szent-Györgyi, J. Isenberg and S. Baird, *Proc. Nat. Acad. Sci. U.S.A.*, **46**, 1444 (1960).
41. R. J. Williams, in *The Enzymes;* P. D. Boyes, H. Lardy, and K. Myrback, (editors) Academic Press, New York, (1959), Second ed., vol. 1, p. 391; A. Ehrenberg and H. Theorell, *Acta Chem. Scand.*, 1193 (1955); L. A. Heppe and R. J. Hilmore, *J. Biol. Chem.*, **198**, 683 (1952).
42. D. R. Storm and D. E. Koshland jr., *Proc. Nat. Acad. Sci. U.S.A.*, **66**, 445 (1970).
43. G. N. Port and W. G. Richards, *Nature*, **231**, 312 (1971).
44. J. C. Bruice, E. Brown and R. Harris, B. Reuben, Q. Milstein and H. E. Cohen, — cf. Discussions in *Nature*, **230**, 206 (1971); M. I. Page and W. P. Jencks, *Proc. Nat. Acad. Sci. U.S.A.*, **68**, 1678, (1971).
45. M. Volkenstein, *J. Theoret. Biol.*, **34**, 193 (1972); *Izvest. Akad. Nauk S.S.S.R.*, Ser. Biol. *(Communic. U.S.S.R. Acad. Sci., Biology Section)*, **6**, 805 (1971).
46. R. Marcus, *J. Chem. Phys.*, **24**, 966 (1956); **26**, 867, 872 (1957); 46 (a) S. Comoroşan, *Biophys.*, **35**, 117 (1968); S. Comoroşan and P. Murgoci. *Bull. Math. Biophys.*, **31**, 629 (1969), **33**, 373 (1971).
47. A. Graffi and H. Bielka, *Probleme der experimentellen Krebsforschung*, Akad. Verlag. Geest & Portig, Leipizig, (1959), Ch. 2 (transl. into Romanian, Acad. Press. Bucharest 1962) (a). E. Boyland in *Carcinogensis, A Ciba Foundation Symposium*, Churchill Ltd, London, (1959) p. 193 (b). C. Heidelberger, ibid. p. 152.
48. E. Huberman, S. Aspiras, Ch. Heidelberger, R. Grover and H. Sims, *Proc. Nat. Acad. Sci. U.S.A.*, **68**, 3195 (1971); D. Dipaulo, R. Donovan and H. Nelson, *idem* 2958; C. Cookson, H. Sims, R. Grover, *Nature New. Biol.*, **234**, 186 (1971).
49. St. S. Nicolau, in: In *Honour of St. G. Nicolau* (in Romanian), Publishing House of the Academy, Bucharest, (1965) p. 489.
50. A. P. Daudel and R. Daudel, *Chemical Carcinogenesis and Molecular Biology*, Interscience, New York, (1966).

51. A. Pullman and B. Pullman, *Advan. Cancer Res.*, **3**, 117, (1955).
52. J. A. Miller and E. C. Miller, *Cancer Res.*, **23**, 229, (1963).
53. A. Heller and B. Pullman, *Cancer Res.*, **19**, 618 (1969); A. Pullman, *Compt. Rend. Soc. Biol.*, **226**, 486, (1968).
54. A. C. Allison and T. Nash, *Nature*, **197**, 788 (1963); A. C. Allison, M. E. Peover and T. C. Gough, *Nature*, **197**, 764 (1964).
55. G. E. Mihailovski and Ia. I. Kozlov, *Biofizika*, (Moskow), **12**, 938 (1967).
56. J. H. Weisburger, E. D. Bergmann and B. Pullman, *Science*, **165**, 417, (1969).
57. T. Hill and M. Morales, *J. Amer. Chem. Soc.*, **73**, 1656 (1951).
58. A. Cammarata, in: *Mol. Orbital Stud. Chem. Pharmacol., Symp.*, (1969); L. B. Kier and B. Lemont (editors), Springer, New York, (1970) p. 156 (a). Y.C. Martin, quoted herein; (b). A. Cammarata, Yan, quoted herein; (c). P. Peradijordi and A. N. Martin, quoted herein.
59. J. Ariens, A. M. Simonis and J. M. Van Rossun, in: *Molecular Pharmacology*, E. J. Ariens (editor), Academic Press, New York, (1964) p. 119.
60. B. Belleau and G. Lacasse, *J. Med. Chem.*, **7**, 768 (1964); B. Belleau, *Ann. N. Y. Acad. Sci.*, **144**, 705 (1967).
61. G. Klopman and R. F. Hudson, *Theoret. Chim. Acta*, **8**, 165 (1967); G. Klopman, *J. Amer. Chem. Soc.*, **90**, 223 (1968).
62. A. Inouye, Y. Shinagawa and Y. Takaishi, *Arch. Intern. Pharmacodyn.*, **144**, 319 (1963).
63. A. Cammarata and R. L. Stein, *J. Med. Chem.*, **11**, 829 (1968).
64. H. Berthod and A. Pullman, *J. Chim. Phys.*, **62**, 942 (1965).
65. J. B. Moffat, *J. Theoret. Biol.*, **40**, 247 (1973).
66. A. Pullman, *Fotschr. Chem. Forsch.*, **31**, 45 (1972).
67. H. C. Keefer, E. J. Condon, I. S. Scarpa and I. M. Klotz, *Proc. Nat. Acad. Sci. U.S.A.*, **69**, 2155 (1972).
68. H. Umeyama, A. Imamura and Ch. Nagata, *J. Theoret. Biol.*, **41**, 485, (1973).
69. J. Kaneti, *J. Theoret. Biol.*, **38**, 169 (1973).
70. R. Gaspar, jr. *Acta Biochim. Biophys.*, **7**, 285 (1972).
71. I. Schwartz, *Studii Cercet. Chimie*, **21**, 721 (1973); J. A. Jhdanov, *Uspehi sovrem. biol.* (Russian), **61**, 187 (1966).
72. R. T. Wedding, C. Hansch and T. R. Fukuta, *Arch. Biochem. Biophys.*, **121**, 9 (1967).

5. Spatial conformation of biomolecules

5.1. Geometric characterisation of biopolymer spatial conformation

The spatial conformation of biopolymers is determined by the orientation (twisting) towards the bonds of which the biomolecular backbone consists. Each monomeric unit contains a number of such bonds and may be characterised by the values of the corresponding angles.

The spatial orientation of the amino acid residues from the polypeptide bond (except for the terminal amino acids), is characterised by three angles, ψ, Φ and ω [1] (Figure 5.1), corresponding to angles of twisting around N—C_α, C_α—C and C—N bonds, respectively. As the C—N bond is partially double owing to conjugation, the ω angle is fixed to either 0° *(trans* configuration) or 180° *(cis* configuration). In polypeptides one encounters the *trans* configuration. In the case of proline, the N—C_α bond belongs to a pyrrolidine ring, and Φ is also fixed to 30°. The Φ, $\psi = 0$ orientation is the corresponding *trans* conformer.

From among the usual secondary structures for polypeptides, the right-handed (R_α) and the left-handed (L_α) α-helical structures are characterised by $\Phi \cong \psi = +120°$ and $-120°$, respectively; the β structures, by $\Phi = 25°$ and $\psi = -25°$ respectively (Figure 5.2). The dimensional characteristics of these secondary structures: for α-helix, one turn of 3,7 amino acid residues, i.e. 5.44 Å; for β-structure the polypeptide chain has a length of 3.63 Å per amino acid residue. The van der Waals thickness of the α-helix is about 21 Å, a period of 18 amino acids (or

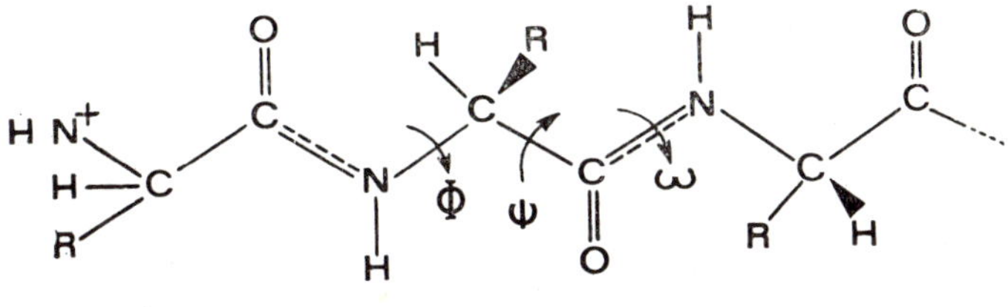

Figure 5.1. Spatial conformation of amino acid residues around bonds from polypeptide backbone.

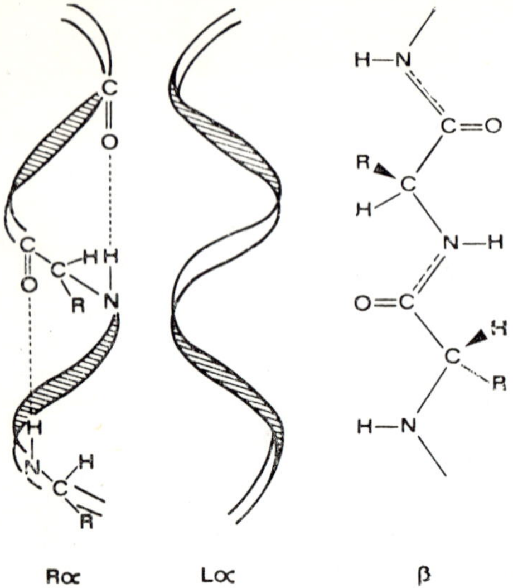

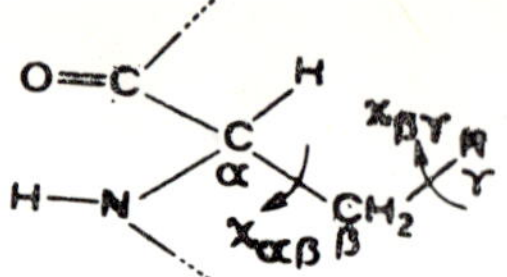

Figure 5.3. Side chain orientation.

Figure 5.2. Ordered secondary structure at polypeptides; R_α = right-handed α-helix; L_α = left-handed; α = helix; β = extended β-structure.

21 Å) along the helix axis. The formation of hydrogen bonds in α-helix and between two β-structured polypeptide sequences, is drawn schematically in Figure 5.2 [2a, 2b]. Another type of structure, sometimes considered as coil, is the reverse turn consisting of four amino acids with 4.8—5.7 Å between the first and last aliphatic C_α atoms and a somewhat longer (<3.2 Å) hydrogen bond between the first carbonyl oxygen and the last imino-group [58].

The side chains are bonded to the C_α atoms of the polypeptide backbone by two CH_2 groups (Figure 5.3). Usually, the R-groups are smaller groups or rigid aromatic rings. Only in the case of some amino acids such as Lys or Arg, do the R-groups additionally contain single bonds which assume distinct orientations. Thus, the spatial orientation of amino acidic side chains may in most cases be characterised by two angles, $\chi_{\alpha\beta}$ and $\chi_{\beta\gamma}$, representing the twisting about the $C_\alpha — C_\beta$ and $C_\beta — C_\gamma$ bonds, respectively.

In case of nucleic acids there are several simple bonds in the biopolymer backbone; thus much more angles per nucleotide residue are required in order to characterise the spatial position of the nucleotides (Figure 5.4). According to nomenclature advanced by Sundaralingam [2], the orientation of the base plane determined by rotation around $C_1 — N_{(2 \text{ or } 9)}$, is characterised by the angle χ. The twisting around the five bonds of the

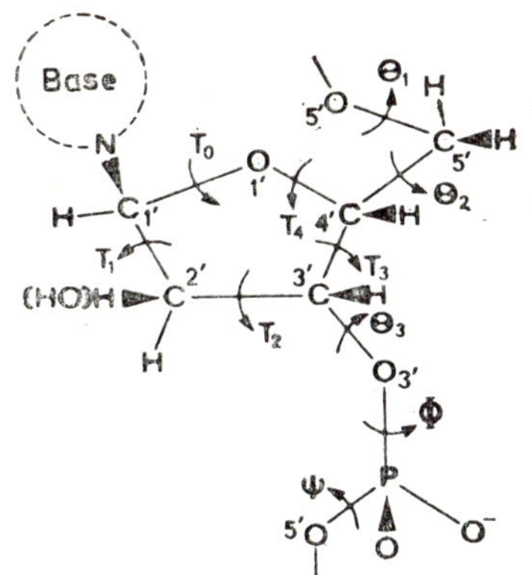

Figure 5.4. Spatial orientation of nucleotide residue.

furanose cycle is characterised by angles T_0, T_1, T_2, T_3, T_4. The phosphate-sugar backbone is characterised by the angles

$$\psi(O_5' - P), \quad \Phi(P - O_3'), \quad \theta_3(O_3' - C_3'), \quad T_3(C_3' - C_4'),$$

$$\theta_2(C_4' - C_5'), \quad \theta_1(C_5' - O_5').$$

Examination of structure deduced from X-ray and steric models for polynucleotides, nucleotides and related phosphoesteric derivatives, has lead to the conclusion that for the helical nucleic acid backbone, the possible conformations are surprisingly limited, and the preferred conformation of the nucleotide unit from the polynucleotide and nucleic acids is the same as that of crystalline nucleotide [2]. A detailed X-ray study of the DNA double helix was perfomed by Wilkins and co-workers [3]. Here again the helix turn is 3.36 Å or 36°, and the van der Waals thickness, *ca.* 20 Å [2a, 2b]. Thus, there are ten base-pairs for a period, with planes perpendicular to the helix axis and almost superposed — by a 36° rotation over each other. The pentosephosphoric chain is at the outward extremity of the helix, and at pH = 7, each phosphate group carries a negative charge, neutralised by cations present in the aqueous solution. The pentose-ring planes are parallel to the longitudinal axis of double helix. The DNA double helix is schematically drawn on Figure 5.4a. The base-pairs — always one purine and one pyrimidine — have equal distances (of *ca.* 11 Å) between the purine and pyrimidine — N, respectively, to which pentoses are bonded (Figure 5.4b).

The distance between the C_1' atoms in the deoxyribose rings of the opposite, antiparallel chains is 11 Å; the length of hydrogen bonds ranges

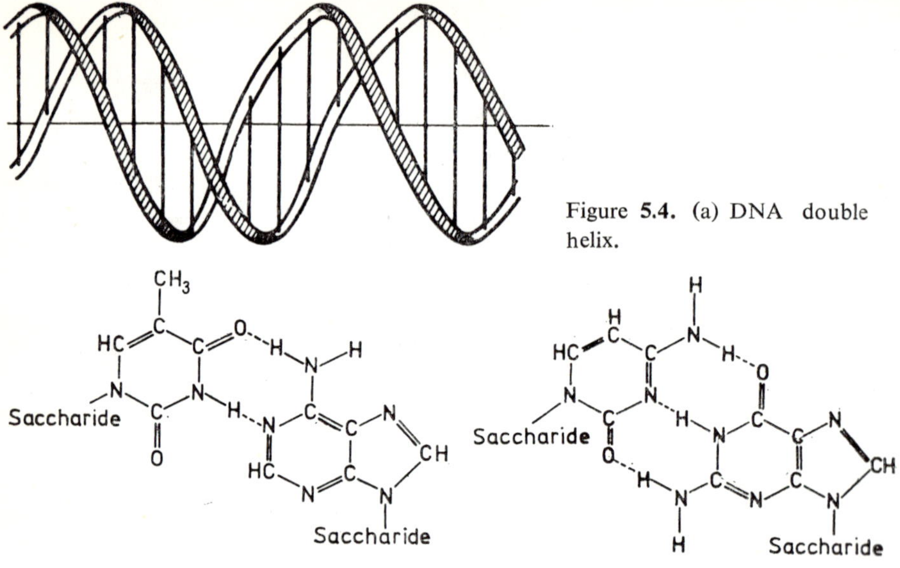

Figure **5.4.** (a) DNA double helix.

Figure **5.4.** (b) Base-pairs of DNA double helix.

from 2.8 to 2.95 Å. The so called grooves in the double helical structure are established as depressions between the negatively charged hydrophilic sugar-phosphate backbones. The wide (major) groove is *ca.* 22 Å accross, measured between phosphates along helix axis, while the shallow (minor) groove is *ca.* 12 Å. The grooves are both *ca.* 7 Å deep, measured from the surface of the enveloping cylinder.

5.2. **Energy calculations regarding biopolymer conformation**

"Energy map" calculations for biopolymers, based on the interaction between atoms, which are not bonded by valence bonds, may be carried out by equations for interaction potentials given in Chap. 3 (section 3.1). Such calculations require an extensive use of computers; they are easier to do in polypeptides, each amino acid residue having the spatial orientation characterised only by two angles, Φ and ψ, than in polynucleotides, whose pentosephosphate backbones each contain five single bonds per monomer around which there may exist several conformational isomers. As expected, such calculations indicate for the twisting around each simple bond (not involved in a ring), three energy minima for twisting, similar to those corresponding to the three rotation conformers around a bond that implies sp^3 hybridised atoms, as in ordinary organic chemistry.

Such sort of energy map calculations for polypeptides were performed by Liquori's group [1], by that of Scheraga [4], by Soviet researchers [5] and by an Indian group [2, 6]. Calculations for nucleic acids have been carried out especially by Indian researchers [2, 6a, 6b]. For example, the method being used by Scheraga [4] for energy-map calculation of the spatial conformation of a polypeptide, considers both interactions between the non-bonded atoms of polypeptide and between these and the water from solution. The standard bond lengths and valence angles were employed and the initial data included some spatial orientations obtained from a first-approximation analysis of X-ray diffraction spectra. Processing these data by computer has lead to a more accurate characterisation of the spatial conformation, that is comparable to the most accurate analysis of X-ray diffraction spectra of polypeptide.

In the polynucleotide chain, the energy-map calculations [6b] show that the monomeric unit can only correspond to a limited set of conformations centred about the values $\chi = 30°$ or $220°$, $\theta_3 = 180°$, $\psi = \pm 60°$ and $\Phi = 210°$ or $70°$, to a large extent independent from the conformation of ribose residue.

Conformational analysis was done by means of the EHT method (*cf.* section 2.4) or semi-empirical potential functions also for medium-sized molecules of biological interest, such as phospholipids [62a], chlorpromazine derivatives [62b], and phenethylamine derivatives [62c]. For example such calculations for phenethylamine derivatives indicate that all psychotropic active molecules have a steric bulge under the plane of the benzene cycle.

5.2.1. *The stereochemic code of Liquori* [1]

On the basis of computations for homopolypeptides, dipeptides and isolated amino acid units, Liquori concludes that the amino acidic residues in polypeptides may exist in one of the five spatial configurations (characterised by Φ_n, ψ_n angles) of low energy. The spatial configuration of an N-peptide may thus be described by a "word" of $N-2$ letters (the amino acids at the chain ends have no fixed configuration) of a five-letter alphabet. Since the spatial structure of the polypeptide is decided by its primary structure this "stereochemic code" would correspond to "translation" of an N-letter word (in the 20 amino acid alphabet) corresponding to the primary structure, into a same length word (in the alphabet of the five energetically favourable spatial configurations) which corresponds to spatial structure.

Liquori's group performed calculations for interaction energies of non-bonded atoms in homopolypeptides, isolated amino acid residues and for one amino acidic residue of a dipeptide, the second being considered in one of the five spatial configurations of low energy. Only $E_{dis} + E_{tors}$ terms were considered *(cf.* section 3.1) for van der Waals interactions and energies of torsion around double bonds. The results of these calculations for polyalanine (i.e. $\psi_2 = \psi_3 = \ldots = \psi_{N-1}$; $\Phi_2 = \Phi_3 = \ldots = \Phi_{N-1}$) and for the isolated alanine residue are shown in Table 5.1. The energetic maps

TABLE 5.1 **Energetically favourable configurations**

	Independent unit			Infinite helix			
	Φ	ψ	V (kcal mol^{-1})		Φ	ψ	V (kcal mol^{-1})
Ra	117°	131°	—0.4	R_α	110°	133°	—3.2
b	25°	— 25°	0.7	β	25°	— 25°	0.4
c	30°	—110°	0.0	γ	30°	—110°	—0.3
d	105°	— 25°	—0.1	δ	105°	— 25°	—0.4
La	—128°	—120°	0.1	L	—128°	—124°	—2.7

show, in both cases, five minima at about the same positions, as seen also in Table 5.1. The notations R_α and L_α refer to α-helixes, β to the extended structure. The energy of regions outside the minima are somewhat higher than +4 kcal mol^{-1} for the isolated unit and +2 kcal mol^{-1} for the infinite helix. For dipeptide, the positions of minima for the considered residue practically do not depend upon the orientation of the second unit. Calculations for other amino acids indicate practically the same minima, no matter what the side chain, —CH_2R. Exceptions are proline and glycine. For proline $\Phi = 30°$ is fixed by the pyrrolidinic cycle, while ψ has two minimum energy values: —12° and +140°. For glycine, the energy map is centrosymmetric (C_α is no longer asymmetric) and very smooth, with low differences between minima and maxima. In all cases the most stable configuration for the peptide group is the *trans* configuration ($\omega = 0$).

The energy map for the hydrogen bond in C=O···H—N< inter-actions was also calculated by Stockmayer's equation (equation (3.9b), section 3.1.2). The energy map contains several minima all of which are in the high energy region of the $E_{dis} + E_{tors}$ map, except for two minima ($\Phi = \pm 135°$, $\psi = \pm 120°$) which fall near the minima for the R_α and L_α configurations. This explains the preference of polypeptides for α-helix conformations.

Calculations for van der Waals and torsional energies were performed for several side chains, with a torsional barrier of 2.8 kcal mol^{-1} for non-terminal C—C simple bonds and 0 kcal mol^{-1} for C—C bonds next to double bonds or aromatic cycles, as suggested by De Coen *et al.* for simple hydrocarbons. Three energy minima are obtained, not far from torsional angles $\chi_{\alpha\beta}$ of 60°, 180° and —60°. The conformation of each bond within the side chain is characterised by torsional angles close to one of these standard values, labelled for hydrocarbon by G, T and G″ respectively.

According to Liquori's hypothesis of the stereochemical code, each amino acidic residue in a polypeptide should be found in one of the five low-energy configurations (Ra, b, c, d, La). Which of these orien-tations is prefered depends on the interaction between amino acid residues, on the minimal energy condition for the whole polypeptide chain (or the whole protein) in solution. In order to test this hypothesis, the configu-rations of the amino acidic residues in the myoglobinic polypeptide chain were compared with the five configurations supposed to be energetically favourable. Myoglobine contains 153 amino acid residues arranged in eight helical segments, two edges and five non helical segments. Precise figures exist for all atomic coordinates and torsional angles of this pro-tein [8]. The 153 amino acid residues are grouped, as orientation, fairly close to the five predicted minima and the statistical analysis gives satis-factory results. An exception is the Gly-80 residue, but, as mentioned, the energy of this amino acid residue depends only slightly on the Φ, ψ angles. An example of this is listed below: the first 11 amino acidic residues in myoglobine and the configurations according to the Liquori's stereochemical code:

Val—Leu—Ser—Glu—Gly—Trp—GlN—Leu—Val—Leu—His—

—dGT—dG—RaGT—Ra—RaGT—RaGG—RaTT—RaGT—RaG—

—RaGG—RaTG—

The idea of this code is not adopted by all workers in this field. Thus, for example, Scheraga in his review on computations concerning protein conformations [4] does not compare the results with the positions predicted by this code. Several works also deal with a more complete description of protein configurations — as α-helix, β-sheet or coil conformations. One must also take into account that the global shape of the polypeptide, of the inter-polypeptide end distance may be seriously affected by small but systematic deviations from the five predicted conformations, an observation made even by Liquori.

5.3. **Protein structure: semi-empirical considerations and experimental data**

There is, at present, a fairly large number of proteins whose primary structures have been elucidated and whose spatial configuration is known. Thus the 1968-edition of *Atlas of Protein Sequence and Structure* [9] lists the amino acid sequence of over 200 proteins, and there are also review papers regarding the overall spatial configuration of proteins [10]. Accurate energy-map calculations, of the type presented in the previous section (5.2), do only succeed with properly predicting the spatial conformation of some short-chain polypeptides; indeed, they fit the known conformations. For this reason, there exists a great number of works attempting to provide empirical rules which enable one to predict spatial conformations from known primary structures. These are especially rules when α-helical and extended β-structures apply; the remaining alternatives (γ and δ, provided that the code given by Liquori is valid indeed) are considered as random structures.

5.3.1. *Semi-empirical considerations on secondary protein structure*

Much work has been done lately on rules predicting secondary structures in proteins. A first step would be gauging the tendency of various amino acidic residues of favouring or disfavouring a certain kind of secondary structure. Thereafter one should find out which combinations of amino acids give a certain type of secondary structure, and finally one should see to what extent the nature of an amino acidic residue exerts an influence on the secondary structure of a remote polypeptide segment. Such rules have been worked out and a recent brief review has been given in Nature [57].

Such rules predict within an accuracy of *ca.* 75—80%, the appearance of a given residue in a certain position of a polypeptide chain to a given secondary structure. The mean values for globular proteins are: 34% of the amino acid residues are in α-helix regions, 17% in β-pleated sheets and 33% in reverse turns [58]. The last type of secondary structure is often considered simply as a coiled structure.

The first attempt to predict secondary structure seems to be the one of Prothero [11]. On the basis of some statistical analysis he predicts a pentapeptidic chain segment to be α-helical if at least three of the five amino acid residues are Ala, Val, Leu or Glu. The rule seems to be verified fairly well by further statistics [13].

Kotelchuck and Scheraga [12] divide the amino acidic residues in the α-helix favouring "*h*" character or the disfavouring "*c*" character. According to them, the "*h*" residues are: Glu, Leu, Met, Trp, Ala, Ile, GlN, Val, Phe, Cys, Arg, Pro; "*c*" residues are: the one preceeding (towards the *N*-end) proline, His, Thr, Asp, Lys, AsN, Ser. Finally, the Gly residue has a neutral "O" character. An α-helix segment is initiated by a succession of four *h*-residues, or four *h*-residues separated by at most a Gly residue. This segment continues to have an α-helical structure, towards the *C*-end, until a succession of at least two *c*-residues appear separated eventually by Gly residues. The first amino acid, next to proline, towards the *N*-end, is always helix-disfavouring (of the *c*-type).

Ptitsyn and Finkelshtein [14] propose *pro-* or *anti*-helix-forming potentials for the amino acids which are smeared uniformly over five residues. Initiation of an α-helix occurs wherever the sum of potentials of five successive residues exceeds a certain "positive" value and is discontinued if this sum decreases below a certain "negative" value. The potential given by each amino acid depends, often, on the direction (towards the *N*- or *C*-end) and, sometimes, conflicts with "*h*" or "*c*" character ascribed by Kotelchuck and Scheraga.

For a β-structure, Ptitsyn and Finkelshtein [14] attribute different values to the potentials of individual amino acidic residues: +50 for Ile, Val and Phe, +25 for Leu, Met and Ala, —25 for Cys, Thr, Tyr, AsN and Ser and —50 for Glu, His, Arg, Asp, Lys and Pro. No "smearing" is considered for those individual potentials. A β-structure is thought to be initiated when the sum of successive potentials exceeds 135 units.

Other figures for individual tendencies of amino-acidic residues to form or break α-helix structure were given by Robson and Pain [15] (IHFinformation, helix forming) based on statistical analysis of pairs of residues. Wu and Kabat [18] have listed the pairs of residues in a triplet together with probabilities of determining a helical configuration for the median residue; however the data available do not cover the whole 20×20 table of possible amino acid pairs.

Concerning the reverse turns, the only statistical information available seems to be that positions 1, 2, 3, and 4 of reverse turns have a strong tendency to be occupied by Asp (19%), Pro (33%), AsN (24%) and Trp (26%), respectively [58].

Table 5.2 lists the frequency of the 20 amino acidic residues in different types of secondary structure regions, according to the statistics

TABLE 5.2 **Frequency of various amino acid residues in different secondary structure regions**

Amino acid	*α-Helix*	*β-Sheet*	*Coil*	*K* coil-helix*
Gly	0.190	0.151	0.660	0.59
Ala	0.522	0.158	0.320	1.10
Val	0.409	0.276	0.315	<1
Leu	0.480	0.209	0.311	1.30
Ile	0.358	0.292	0.349	—
Phe	0.402	0.207	0.390	1.28
His	0.446	0.121	0.432	>1
Tyr	0.220	0.240	0.540	<1
Trp	0.409	0.205	0.386	—
Pro	0.212	0.165	0.624	⩽1
Cys	0.278	0.222	0.500	<1
Met	0.429	0.286	0.286	>1
Ser	0.282	0.149	0.569	<1
Thr	0.295	0.237	0.148	—
Glu	0.549	0.044	0.409	1.28
Asp	0.351	0.135	0.514	<1
GlN	0.421	0.253	0.326	<1.28
AsN	0.263	0.120	0.617	—
Lys	0.383	0.109	0.509	1.14
Arg	0.282	0.167	0.551	—
Mean values:	0.359	0.180	0.461	

* Equilibrium constant for the coil-helix transition in synthetic polypeptides nearly physiological conditions; water solution (25°C).

of Chou and Fasman [59] for proteins containing 2437 amino acidic residues of known sequence and conformation. Leu is strongly helix-forming and appears often in inner helical cores of proteins which suggests the major role of this amino acid, as a nucleation center in the evolution of protein molecules. Gly is the most frequent amino acid in the coil-region. His is often found at the helix-coil boundaries, while Pro is often the last amino acid in an α-helical region towards the *N*-end of the polypeptide.

Equilibrium constants for coil-helix transitions in synthetic polypeptides (in conditions close to physiological ones) are also listed in Table 5.2, according to Ptitsyn [60]. This author also lists frequencies of finding amino acid residues in α-helical regions, from a statistical study of nine globular proteins with results closely resembling those of Chou and Fasman.

The conformation of a certain residue also depends on its neighbours near or far, that is on the so called short-range and long-range interactions. A logical analysis indicates that the most important influence is that of the neighbours separated by one amino acidic residue. The influence decreases rather slowly with increasing distance; the influence of neighbours separated by at most six or at least seven amino acidic residues seems of almost equal importance. Predictions of secondary structures, taking also into account the influence of neighbours separated by 0, 1, 2, ..., 6 residues do not seem to exceed 80—90% precision [61a]. This seems to conflict with the result of Ponnuswamy *et al.* [61b] who have computed the dihedral angles which minimize the configurational energies of oligopeptides, and claim that if all nonapeptides are considered, the configuration of lowest energy is, in most cases, the one given by X-ray diffraction for lysosine.

The tendency of amino acidic residues to favour or disfavour a certain secondary structure, may be particularly well explained by interactions of the side chain with the polypeptidic backbone, within a dipeptide range. The "regular" structures (α-helix, β-sheets) lower the energy of the peptidic backbone by hydrogen bond formation but generally increase the energy of the side chains, due to difficulties in accommodating them in the regular spatial structure. The polar or charged groups of the side chains, on the other hand, may tend to form hydrogen bonds with either the peptidic backbone or with water [16, 55]. Long-range interactions may also play a rôle. By these means, Vorobiev and Finkelshtein [17] have explained the secondary structure of histone f2a-1 at neutral pH and medium ionic strength. The non-helical regions contain almost only bibasic amino acids; within α-helical regions, positively charged side chains are often three to four residues apart from negatively charged ones.

5.3.2. *The overall spatial conformation of proteins*

According to studies of mild denaturation on some enzymes whose activity is recovered following renaturation, this is determined by the primary structure of protein chains. The same occurs with the reduction-reoxidation of disulphide bonds [9]. The spatial conformation of the protein in aqueous solution is also retained in crystals obtained from the solution [20]. In general, the forms assumed by proteins in solution are fairly diverse, yet the most characteristic feature is the packing of the hydrophobic amino acid residues in the inner core or 'globe', forming a sort of semisolid "oil droplet" [21]. According to Lim and Ptitsyn [54], the hydrocarbon chains of aliphatic amino acids and of phenylalanine may be packed according to a diamond-like tetrahedral lattice ensuring a maximum number of van der Waals contacts between the atoms of these groups. In this packing, the most important rôle is played by the interaction of each side chain with that of the amino acidic residue at the right and left, respectively. The X-ray structural analysis of myoglobin agrees well, within experimental error, with those tetrahedral packings — from among all possible tetrahedral alternatives — which, according to calculations, possess the most negative energy of chain interaction. Thus, the spatial structure of proteins is probably the most thermodynamically stable structure corresponding to a certain primary structure and one may speak of a "stereochemical code", though not necessarily *stricto sensu*, of the five permitted conformations of amino acid residues (Liquori [1]). Even this strictest sense is likely enough: X-ray analysis shows that monomer units (excepting Gly- and Pro-residues) of protein chains have a conformation close to Liquori's Ra- and β- and, occasionally La-configurations [22]. From among the 150 amino acidic residues of the 50 myoglobins studied so far, only eight residues are homologous.

The question arises as to the way in which the polypeptide chain reaches its 'natural' spatial structure. De Coen [23], based on computer simulation (for conformational energy calculations), claims that as the synthesis of the polypeptide chain on the ribosome proceeds, at a length of about six residues, the most thermodynamically stable configuration is formed and the remaining part of the chain — as its synthesis proceeds — coils on this starting core. Under these circumstances it would be more difficult to explain the renaturation of a completely uncoiled polypeptide. Regarding random evolution towards the most thermodynamically stable configuration, such a sequential process would require a time longer than the duration of denaturation [24]. The mechanism proposed by Scheraga appears to be the most likely [25]: on the basis of nearest-neighbour interactions, certain amino acidic segments strongly tend to adopt — and indeed, often do — a certain spatial conformation. These short regions are stabilised by interactions with remote neighbours from other polypeptide portions. By means of this coiling around initial nuclei the polypeptide will, quite rapidly, gain a final form corresponding to a minimum of free enthalpy G, yet not necessarily the lowest minimum. Calculations for the auto-organisation of the myoglobin molecule were carried out by Ptitsyn [63].

The process, which starts up with fluctuating secondary structure regions, forming around "crystallisation centres" requires *ca.* 1×10^{-4} s for the nucleation process and 3×10^{-4} s for growth and nucleation. The stabilisation energy for crystallisation in myoglobin was calculated and found to be —90 kcal mol^{-1} [63]. Lim [64a] and Finkelshtein [64b] imagined some rules relating the whole shape of globular proteins to secondary structure, *via* long-range interactions between amino acidic side chains, the packing of hydrophobic side chains and a statistical analysis concerning the primary and secondary structure of 18 proteins of known structure. They concluded that there is a feedback between primary and secondary structure in globular proteins, in the sense that the existing spatial conformations are the most favoured ones both by the short-range and the long-range interactions. It appears that during evolution those conformations were selected which have been favoured by both types of interactions and thus have been more stable against irreversible denaturation [64c].

Two very important indices in the determination of the overall form and of the tendency of association, are the mean hydrophobicity, $\overline{H\Phi}$ (the sum of Tanford hydrophobicity increments, Table 3.11, section 3.3.7, divided by the number of amino acids in the protein) and the mean fractionary charge, $\overline{FC}$ [26]. Proteins with many polar groups, have in solution a stretched form; those with high $\overline{H\Phi}$, a spherical form, and, as a rule, a marked tendency to association. A study of the crystallographic

spatial structure of several proteins reveals that almost all of the ionic side chains are completely exposed to the solvent, but a large fraction — *ca.* 50% — of the non-polar residues are also available to interact with the surrounding aqueous solution [27]. A $\overline{H\Phi}$ of at least 0.91 kcal mol^{-1} is required for a globular structure to appear: all globular proteins studied have $\overline{H\Phi} = 1.00 \pm 0.10$ kcal mol^{-1}. The association tendency of proteins is higher the lower the $\overline{FC}$. Thus myoglobin ($\overline{FC} = 0.36$) and the cytochromes ($\overline{FC} = 0.33$) are monomers in solution, while α- and β-globins have $\overline{FC} = 0.26$ and they tetramerise ($\alpha_2\beta_2$ haem$_4$) in solution. The heavy and light chains of immunoglobulins have an $\overline{H\Phi}$-value close to 1.0 kcal mol^{-1} and an average $\overline{FC} = 0.19$, which explains their tendency to associate (tetramerisation) in solution [26].

Forcing the polypeptide chain into a unique spatial conformation— from many possible — entails a loss of configurational entropy, which is, per mole of amino acidic residue (two simple bonds each with three approximately equi-energetic twisting positions):

$-S_{conf} = N_A k \ln W = R \ln 3^2 = 4.1$ cal deg^{-1} mol^{-1} of aa-residue. The number of microstates corresponding to random orientation per aa residue is $W = 3^2$. This loss of configurational entropy should be compensated by other effects. By calculating the sum of hydrophobic increments of the component amino acids, $\Sigma H\Phi$, one finds that it approximately compensates the $-T\Delta S_{conf}$ term for proteins without disulphide bonds. In the presence of disulphide bonds, the $-T\Delta S_{conf}$ term is larger than $\Sigma H\Phi$; the excess of configurational entropy loss is balanced by the energy of the disulphide bonds [26]. In Table 5.3 are given data discussed

TABLE 5.3 **Values of $\Sigma H\Phi$ and $T\Delta S_{conf}$ (300 °K) for some globular proteins**

Protein	*No. of aa residues*	σ, *no. of SS bonds*	$-\Sigma H\Phi$ (kcal mol^{-1})	$-T\Delta S_{conf}$ (kcal mol^{-1})	$\frac{\Sigma H\Phi - T\Delta S}{\sigma}$
Myoglobin	153	0	173	184	
Cytochrome *c*	104	0	121	125	
TMV envelope protein	158	0	172	190	
Subtilysine	275	0	265	330	
IgG light chains	215	2	219	258	19.5
Ribonuclease	124	4	109	149	10
Lisozyme (hen)	127	4	127	155	7
Chymotrypsinogen A	245	5	252	294	8.5

above (according to [26], Table 6); in no case does the $\Sigma H\Phi\text{-}T\Delta S_{conf}$ difference per disulphide bond exceed the energy of the disulphide bond (of 50 kcal mol^{-1}; Table 1.1, Chap. 1).

Another characteristic of specific protein volume in solution is the fact that this volume exceeds only by a little the overall sum of specific volumes of amino acidic residues [28]. The additional volume, also named the conformational volume, may reach as much as *ca.* 3% of the total volume, or 0.02 cm^3 g^{-1} of amino acidic residue (average molecular weight = 125); this excess volume corresponds to 2.5 cm^3 surfaces "towards vacuum". Separated by a distance of 1—1.5 Å these surfaces would be of *ca.* 2×10^8 cm^2. For a surface tension of σ = 25 dynes cm^{-1}, the corresponding surface energy (*cf.* section 3.3.3) would be of 5×10^9 erg, or *ca.* 0.12 kcal mol^{-1} per aa residue. Another way of evaluating these "empty" spaces, or, perhaps more accurately, the extensibility of hydrophobic centres of proteins, is by measuring the solubilisation of hydrocarbons in aqueous protein solutions. Thus, for serum albumin (1% solution, 20 °C, isoelectric pH), this "extensibility" represents *ca.* 6% (*ca.* 7 000 Å^3) of the total protein volume [31].

The structure of globular proteins is probably not completely rigid, completely "crystallised" in solutions. Some features of the n.m.r. spectra of proteins as the pancreatic trypsine inhibitor indicate that there is some permanent degree of fluctuation within the protein structure, under physiological conditions [65].

5.3.3. *Spatial conformation of some proteins*

The protein structure depends on several structural and environmental factors. For example, the secondary structure of some polymers of H(Lys)$_m$—Ala$_n$—(Lys)$_m$OH type, where $n \gg m$ in aqueous solution, is entirely non-helical for $n = 10$, but helical for $n \geqslant 160$. Or, the poly-*L*-methionine-*S*-methylsulphonate at ionic strengths of $\mu \leqslant 0.015$ M is randomly coiled in the presence of Cl^- and Br^- ions; it assumes an antiparallel β-structure in the presence of I^- and CNS^- ions, and an α-helix in the presence of ClO_4^- [29]. Poly-Lys in water is α-helical at basic pH (the ε-NH_2 groups are neutral), yet it becomes randomly coiled if the ε-NH_2 groups are positively charged [4]. In Table 5.4 are given the α-helical percentages of some proteins in aqueous solutions close to physiological conditions (according to [30]).

TABLE 5.4 α-**helicity (per cents) for some proteins in aqueous solution**

paramyosin	96%
tropomyosin	86—88%
meromyosin (light fraction)	87%
myoglobin	70%
myosin	60%
bovine serum albumin	46—55%
meromyosin (heavy fraction)	52%
insulin	38%
fibrinogen	33%
egg-white albumin	31%
pepsin	31%
lysozyme	29%
ribonuclease	16%
histones	20%
globulin (H-chain)	15%

A well-studied protein is myoglobin (according to [32a,b]). This protein consisting of 153 amino acids with a molecular weight = 175 000 daltons, has a volume $V = 2.2 \times 10^4$ Å^3, dimensions of 25 Å × 35 Å × × 35 Å. In the native state it may exist in at least two separate conformations, slightly different with respect to structure and stability, but implying important variations from a functional standpoint.

Ribonuclease has been studied extensively [33a,b,c]. Its overall structure is schematically given in Figure 5.5. The molecule has an approximately "kidney-like" shape, and dimensions of *ca.* 38 Å × 28 Å × 22 Å, with a deep "hollow" on one side. The enzymic activity was studied by cleaving various portions of the peptide chain and measuring the residual enzymic activity. The active centre of the enzyme contains His-residues at positions 12 and 119, and Lys at 7 and 14. Thus, the actual spatial structure of the molecule implies a "folding" which brings together spatially these relatively remote (on the polypeptide chain) amino acidic residues.

The structure of antibody proteins — a typical representative is the immunoglobulin IgG — was elucidated by Edelman [34]. The proteins

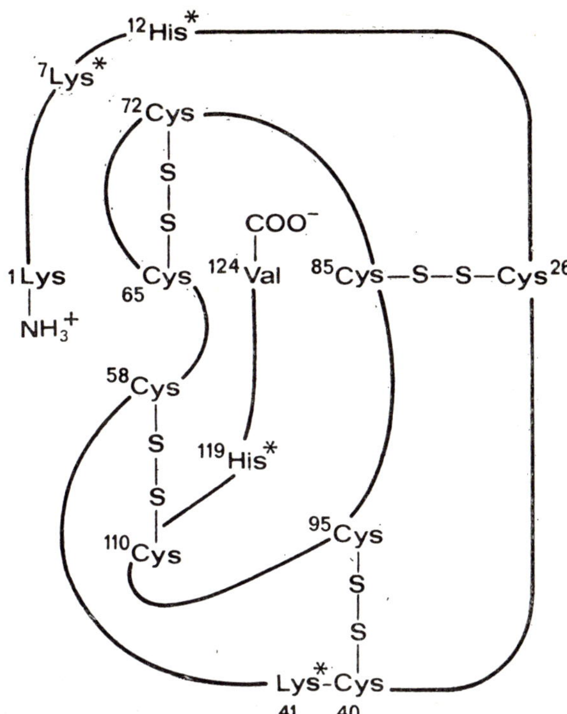

Figure 5.5. Schematic spatial structure of ribonuclease. The amino acids labelled by an asterisk belong to the active centre of the enzyme.

consist of two heavy polypeptide chains *H* of 450 amino acids, and of two light chains *L* of 213—214 amino acids (Figure 5.6). Both types of chain do also have variable regions, with respect to amino acid composition (V_A and V_L, respectively), and constant regions for a given organism. Every IgG molecule has two combining sites for antigens, located in between the variable regions of each *H*- and *L*-chain. These sites contain several Trp residues [34b] and have a length of some 10—15 amino acid residues [34e]. The *H*- and *L*-chains are held together in the complex H_2L_2 protein by both disulphide and hydrophobic bonds, as seen in Figure 5.6, which features the results of some X-ray studies [34c,d].

5.3.4. *Quaternary structure and allosterism.*

Many proteins are formed from several polypeptide chains. Such proteins are also named oligomeric proteins [35], in the sense that they generally contain a relatively small number of identical units, protomers,

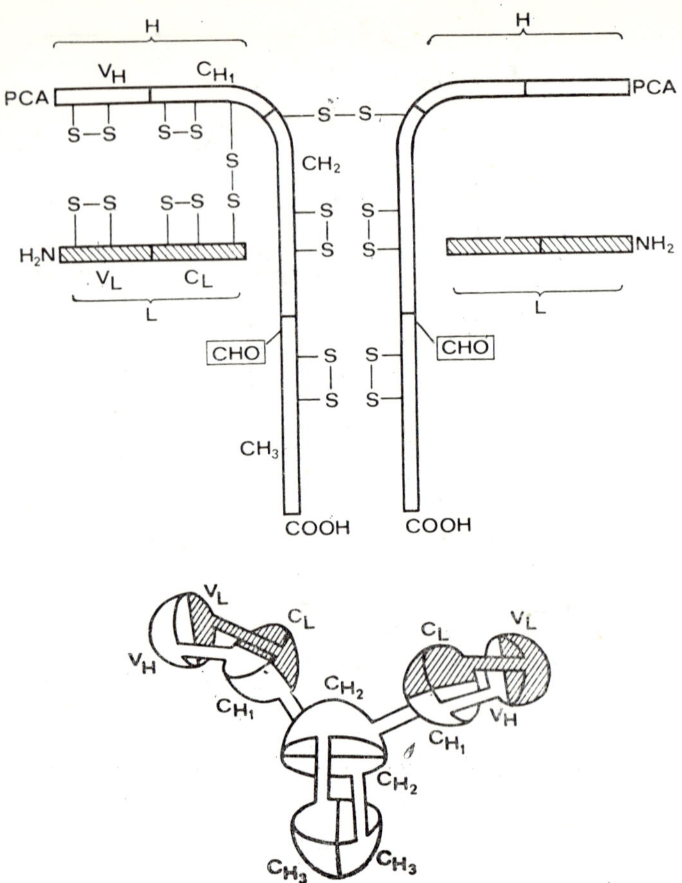

Figure **5.6.** Spatial structure of immunoglobulin, IgG (H_2L_2) (according to [34a] and [34c,d].

V_H, V_L variable regions; C_L, $C_{H_{1-3}}$ constant regions in L- and H- chain; PCA = pyrrolidine carboxylic acid; CHO = carbohydrate.

which in their turn may be formed from one or more monomers. The monomer is, in general, a polypeptide chain. Most often, oligomers consist of two or four polypeptide chains, yet sometimes larger numbers are encountered — such as ten chains (of 19 000 dalton, each) — in histidine carboxylase, or 2130 chains (each of 17 500 dalton) in the TMV (tobacco mosaic virus) particle. The molecular weight of these polypeptide chains varies within wide range whose known extreme limits are 5733 dalton and 4 100 000 dalton [36].

Allosterism occurs in the interaction of a protein with several identical or different ligands, and consists of the fact that even though the sites that bond the ligands are equivalent, the affinitiy of a site towards a ligand depends upon the position of the other sites (occupied or not by ligands) [35]. Most of allosteric proteins known are oligomeric; the allosteric effects are due to alteration of quaternary structure, i.e. to binding among subunits. These allosteric effects may be homotropic or heterotropic, according to whether the ligands are identical or not, respectively. The heterotropic effects seem always to be accompanied by homotropic ones; however, while the heterotropic effects may be either positive or negative (that is co-operative or antagonistic), the homotropic ones seem to be always co-operative (positive), in the sense of increasing affinity for the next ligand. The problem of allosteric interaction, with a ligand L, of two allosteric forms A_i and B_i ($i = 0, 1, \ldots, n$ denotes the number of ligands bonded to the oligomer in the corresponding allosteric form) can be treated by means of the theory of thermodynamic equilibria [35]. Let us consider the successive equilibria:

$$\mathrm{A}_0 \rightleftharpoons \mathrm{B}_0 \tag{5.1a}$$

$$A_{i-1} + \mathrm{L} \rightleftharpoons \mathrm{A}_i; \quad \mathrm{B}_{i-1} + \mathrm{L} \rightleftharpoons \mathrm{B}_i \quad i = 1, 2, \ldots, n \tag{5.1b}$$

and the equations describing these equilibria:

$$\frac{[B_0]}{[A_0]} = K; \quad \frac{[A_i]}{[A_{i-1}]} = \frac{n-i}{i} \cdot \frac{[L]}{K_\mathrm{A}}; \quad \frac{[B_i]}{[B_{i-1}]} = \frac{n-i}{i} \cdot \frac{[L]}{K_\mathrm{B}} \tag{5.2}$$

If one introduces the symbols:

$$\frac{[L]}{K_\mathrm{A}} = \alpha, \quad \frac{K_\mathrm{A}}{K_\mathrm{B}} = \gamma,$$

one obtains for fraction $\bar{A}$ of the protein in oligomeric form A:

$$\bar{A} = \frac{(1+\alpha)^n}{K(1+\gamma\alpha)^n + (1+\alpha)^n}, \tag{5.3}$$

and for the average fraction $\bar{\lambda}$ of the sites occupied by ligands:

$$\bar{\lambda} = \frac{K\gamma\alpha(1+\gamma\alpha)^{n-1} + \alpha(1+\alpha)^{n-1}}{K(1+\gamma\alpha)^n + (1+\alpha)^n}. \tag{5.4}$$

The assumption was made that the overall successive binding constants depend only on the ratio of the $n-i$ number of free sites over the number i of sites bonded on oligomer A_i or B_i. If $K_A = K_B$ and $[L] \ll K_A$, one obtains for $\bar{\lambda}$ the Michaelis-Menten equation:

$$\bar{\lambda} \cong \alpha \frac{\alpha}{1+\alpha} = \frac{[L]}{K_A + [L]}. \tag{5.4a}$$

Strong co-operative effects are obtained for concentrations of $[L] \gg K_A$ and for small values of γ. As a measure of these, one may use with homotropic effects Hill's approximation:

$$\bar{\lambda} = \frac{[L]^{\bar{n}}}{Q + [L]^{\bar{n}}}. \tag{5.4b}$$

For example, for dCMP — deaminase, for the substrate — dCMP —, $\bar{n} = 2.0$. The dCMP activator rises the deamination velocity according to equation (5.4b), with $\bar{n} = 2.0$ for concentrations of about 6.7×10^{-5}M, while the inhibitor dTTP, for concentrations of 4×10^{-4} M, decreases this velocity according to equation (5.4b), with $n = 3.4$.

Monod and his co-workers [35] have been able to draw some interesting conclusions concerning the geometry of the bonding between protomers in oligomers. In allosteric proteins all the proteins occupy equivalent positions, thus implying the existence of at least one axis of symmetry. The symmetry of each set of stereospecific receptors for ligands is identical to the symmetry of the molecule. At least two states, different with respect to hindrances imposed on the protomeric configuration and to the affinity of stereospecific sites towards ligands, are reversibly accessible to allosteric oligomers. All the allosteric states of oligomer have the same symmetry. Although the association of protomers in oligomers does not seem to imply covalent bonds, the specificity of association is extremely high: the monomers of a normal protein recognise their own identical partners and reassociate even at very low concentrations and in the presence of other proteins.

5.4. Spatial conformation of nucleic acids

The crystalline DNA molecule may exist in three forms, namely A, B and C [37]. From these, form B exists at high humidity, as well as in solution; it is the double helix usually refered to. It has generally been assumed that DNA in aqueous solution exists primarily in the B-conformation. In the fibre (gel) state the B-form is easily converted into A- or C-form, by varying humidity or counter ions. In the A-form the base pairs are tilted about 20° from the perpendicular to the helix axis and are displaced outward from this axis. There are eleven base pairs per turn in this structure with a rise per residue of 2.82 Å. Double stranded RNA at 100% relative humidity greatly resembles the A-form. The C-form resembles B, but has only 9.3 base pairs per turn and a rise per residue of 3.32 Å [66].

In molecules of RNA and synthetic polyribonucleotides, the double helix-forming tendency is far less evident than in polydeoxyribonucleotides [37]. This might account for the genetic information being enconded (following selection) in DNA and not RNA; meanwhile substitution of thymine for uracil might be due to the higher photochemical stability of the former. Indeed, thymine does form a dimer upon U.V.-irradiation, yet this is efficiently reversed on irradiation at higher wave lengths [38].

The structure of RNA in solution or crystals and the degree of stacking (parallel superposed arrangement) of bases may be studied by physico-chemical methods, such as studies of solution viscosity or hyperchromic variations [37]. At high salt concentrations and not too high temperatures, the RNA molecules have, in general, quite a compact structure, with small double-helical portions which melt on increasing temperature. At low ionic strengths and temperatures a rod-like structure emerges having a marked rigidity, favoured by traces of polyvalent cations.

An extensively studied structure is that of tRNA [39], a relatively small molecule, with molecular weight of 25 000—30 000 dalton, degree of polymerisation of *ca.* 80, against the degree of polymerisation of mRNA of *ca.* 10^3, corresponding to about 3×10^5 dalton, and 10^7—10^8 dalton for DNA. According to X-ray diffraction studies it seems that all of tRNA species have similar spatial structure — a sort of clover-leaf, folded about a stretched double-helical core (*cf.* Figure 5.7).

The same nucleotides (denoted by capital lettering — the base symbol) or usually the same nucleotide (labelled by a bar above base symbol) are always met with in many positions. In the tRNA chain (several of which have already been sequenced) one encounters unusual bases, which apparently do not originate in gene transcription, but in modifications occuring

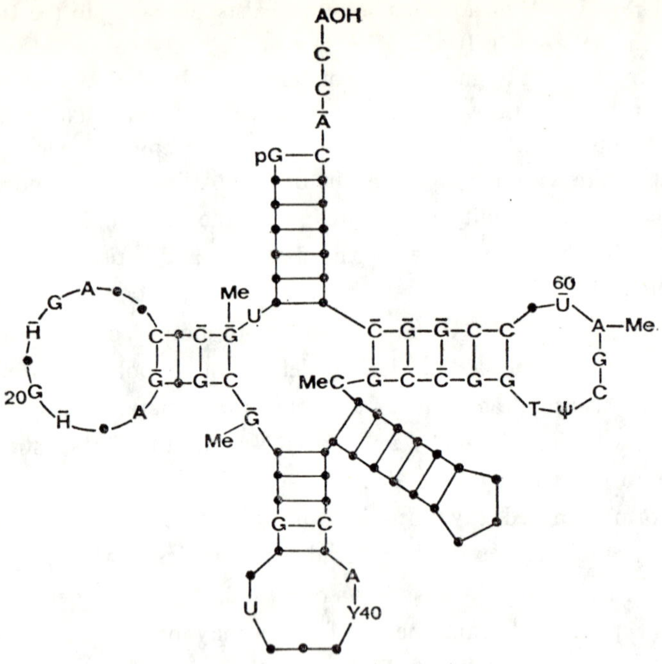

Figure 5.7. The clover-leaf structure of tRNA (according to Phillips [39c]).

thereafter in the transcription products. In Figure 5.7, H stands for dihydrouracyl, ψ and γ for other odd-bases, while Me represents methyl groups for methylated bases. The anticodon is represented at the bottom opposed to 5′-terminal adenine carrying the activated amino acid. In the stretched-folded form (in which tRNA does actually exist in solution) of clover-leaf, too, the anticodon and activated amino acid are on the opposite ends of the longitudinal axis.

With respect to mesRNA structures, studies on the nucleotide sequence of the gene coding for the bacteriophage MS2 coat protein [67a] suggest that the codons for amino acids, especially for the essential ones, tend to avoid regions with a secondary structure [67b]. Primary structures of mesRNA molecules seem not to contain, at least in the coding sequence, four or more successive bases able to find antiparallel complementarity on the same messenger [67c]. These peculiarities may be the result of natural selection for more reliable translation.

5.4.1. *Semi-empirical considerations on nucleic acid secondary structure*

The DNA molecules, by virtue of native existence of two complementary strands, have a very marked tendency to yield a double helix. Yet, in this case too, more recent X-ray diffraction work on some gels or concentrated solutions of DNA, poly-(A + T) or poly-(AT + TA), have shown that there appear certain differences from classical B-form double helix [40]. Namely, with increasing A + T percentage in DNA-composition, between 56% and 67% a co-operative transition takes place to a helix — having a 10% larger turn than DNA poorer in A + T.

The RNA molecules — usually existing as a single strand, without the complementary one — offer a wider variety of structures, ranging from random-coiled strands to a helix with stacked bases. The tendency to form a helix with stacked bases characterises the purines [41a]: according to optical criteria, poly-rA is a stacked helix, while poly-rU, poly-drU and apurinic DNA (i.e. with 3/4 of pyrimidines remaining not flanked by other bases) have a random-coil structure. This stacking also exists in the uridine dinucleotide and it does not disappear with hydrogenation of one of the bases [48b]; therefore, for dimers the hydrophobic interaction suffices for stacking in aqueous solution, even though not augmented by aromatic interaction. It is characteristic to deoxyribonucleotide oligomers to exhibit a tendency of associating into double helix stronger than that of ribonucleotide oligomers [41d]. Thus $(dA)_n + (dT)_n$ yield complexes with paired bases for $n \geqslant 3$, while $(rA)_n + (rU)_n$ do so only for $n \geqslant 7$.

Rules were issued by Uhlenbeck *et al.* [42] for estimating secondary structure in ribonucleic acids, as a function of their primary structure, based on thermodynamic and configurational entropy considerations and on measurements of "melting" equilibria of the helical segments. These rules are meant for $T = 25\ °C$, average ionic strengths of $\mu \cong 1$ M and pH = 7. There are approximate equations for calculating ΔG of the transition from random-coil to a partially double helix also containing non-paired regions. The various forms of partially paired strands, which may arise at the closing of a RNA-chain segment, are shown in Figure 5.8. The formation of paired regions leads to a free enthalpy gain of —2.4 kcal mol^{-1} per GC-pair, of —1.2 kcal mol^{-1} per AU-pair, and of *ca.* 0 per GU-pair. The formation of closed regions leads to a loss in configurational entropy, $-T\Delta S_{conf}$, of +6 kcal mol^{-1} for the HL-form with

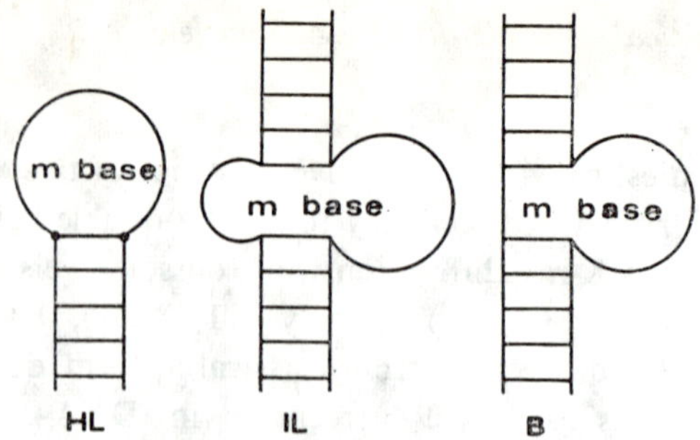

Figure **5.8.** Forms of RNA chain segment with partially double helical structure.
HL = hair pin loop; IL = internal loop; B = "bulge".

$m = 3$ ($m < 3$ is geometrically impossible), of *ca.* +7 kcal mol^{-1} for m between 4 and 7, and of +8.5 kcal mol^{-1} for $m \geqslant 8$. The IL form appears for at least two bases with $-T\Delta S_{conf} \cong +5$ kcal mol^{-1}; this becomes *ca.* +6 kcal mol^{-1} for $m = 3$; *ca.* +7 kcal mol^{-1} for m between 4 and 7, and *ca.* +8.5 kcal mol^{-1} for $m \geqslant 8$. For B-form, $-T\Delta S_{conf}$ is *ca.* +2.5, kcal mol^{-1} for $m = 1$ or 2; *ca.* 3.5 kcal mol^{-1} for $m = 3$; +6 kcal mol^{-1} for m between 4 and 15 and about 7 kcal mol^{-1} for $m \geqslant 16$. In addition, the closing of a m-polynucleotide segment brings about a loss of initiation ΔG of 2.3 RT [3 + log (m + 1)]. Finally, the closure of a m-nucleotide random segment and forming of λ loop forms and β bulge forms (of $m_1 m_2 \ldots m_\lambda$ bases each, in loops 1, 2, ... λ and of $m_1, m_2 \ldots m_\beta$ bases each in bulges 1, 2 ... β) with formation of N_{AU} pairs of AU bases and N_{GC} pairs of GC bases, entails a ΔG change (in kcal mol^{-1}, at 25 °C, pH = 7, μ = 1M) of:

$$\Delta G = -1.2N_{AU} - 2.4N_{GC} + 5.9\,\lambda + 4.1\,\beta +$$

$$+ 2\left[\sum_{i=1}^{\lambda} \log(m_i + 1) + \sum_{j=1}^{\beta} \log(m_j + 1)\right]$$

Assumption was made that all m_i, $m_j \geqslant 4$. For lower (μ=0.2 M) ionic strengths, the base-pair stabilisations are smaller, namely = —1.0 kcal mol^{-1} and —2.2 kcal mol^{-1} for AU- and GC-pairs, respectively. This equation has been applied to calculations of relative G values for different hypothetical conformations of some tRNA molecules, those indicated by X-ray studies included; the lowest G values have been obtained for actual configurations which is, indeed, a verification. Among these calculations of nucleic acids conformations, we mention those of Delisi and Crothers [56]; use has been made of an improved version of the scheme of Uhlenbeck and his co-workers. Taking into account the uncertainties of even the improved version, Delisi and Crothers have considered that a spatial structure calculated for a tRNA may be taken for granted if the remaining structures have calculated G values which are at least 5 kcal mol^{-1} more positive.

Some recent measurements on oligonucleotides [68a,b] have yielded the figures listed in Table 5.5, for loop formation as a function of base composition.

TABLE 5.5 **Free enthalpies (ΔG, kcal $mole^{-1}$ at 25 °C) for loop formation in RNAs versus base composition**

Loop size	*Frontier base pairs*	ΔG (kcal mol^{-1})
2	AU ... AU	+1.8
2	GC ... GC	+0.1
4	GC ... GC	+1.6
6	GC ... GC	+2.5
Opening of central loops (the central base pair dissociated)		
	GGG CCC	+7.5
	GGA CCU	+6.75
	AGG UCC	+6.75
	AGA UCU	+6.0
	AAA UUU	+4.0
	AAG UUC	+4.15
	GAA CUU	+4.15
	GAG CUC	+4.3

5.5. Conformational transitions

The transition from one spatial conformation to another has the features of a second-order transition—it proceeds practically completely in a relatively short interval of variation of an external parameter (e.g. temperature). Quantitatively, within a process, e.g. transition of protein from a random coil structure to an α-helix, the building up of the first α-helical loop requires a more positive ΔG than the subsequent steps (*cf.*, e.g. [2b] or section 1.4). The theory of this transition derives from a model given by Ising (*cf.* for example [43]), which considers only the interaction between the nearest-neighbours on the polymer chain. This model is charaterised by two parameters, s and σ, in case of homopolymers. If ΔG_c is the free enthalpy for "expanding" the ordered structure on polymer chain with a monomer unit, and ΔG_{I} the free enthalpy of formation of the first core of ordered structure, then:

$$s \equiv e^{-\frac{\Delta G_c}{RT}}; \quad \sigma \equiv e^{-\frac{1}{RT}(\Delta G_{\mathrm{I}} - \Delta G_c)} \tag{5.6}$$

Alternatively, if u_j and v_j are the statistical weights for a random coil segment of j-monomers and an ordered segment of j-monomers, respectively [1] [44], one obtains:

$$s = \lim_{j\to\infty} \left(\frac{v_{j+1}}{u_{j+1}} \cdot \frac{u_j}{v_j} \right) . \because . \sigma = s^2 \lim_{j\to\infty} \left(\frac{v_j}{s^j u_j} \right) \cdot \tag{5.6a}$$

When this theory is applied, especially to short-ordered segments (α-helical proteins), one cannot account for the characteristics of reversible denaturation-renaturation processes. Actually, a nucleation process, the formation of an ordered structure within small segments (with amino acid residues which favour especially the structure under discussion) takes place at first. These regions extend and are subsequently stabilised, in the folding process by interactions between various segments of the polymer chain and, between remote monomers (along the chain) [1]. The model is relatively fairly suitable for homopolymers. For heteropolymers other theoretical approaches have been developed, e.g. based on the method of successive quasi-stationary reactions [45a], or specifically for other types of transitions, e.g. random coil-β structure [45b].

5.5.1. *Thermodynamic parameters for conformational transitions*

The conformational transitions take place in the range defined about parameter $s = 1$ or $\Delta G_c = 0$. If there is a quasi-melting, one may define a transformation temperature T_t, about which the conformational transition occurs:

$$\Delta G_c \equiv \Delta H_c - T_t \Delta S_c = 0 \therefore T_t = \frac{\Delta H_c}{\Delta S_c}, \tag{5.7}$$

where ΔH_c and ΔS_c are the enthalpies and entropies corresponding to ΔG_c. At the usual temperatures (300 K), the ΔG_c values for α-helix-coil transitions of proteins are of the order of 0.1 kcal mol^{-1} [30], and for nucleic acid double helix-random coil transitions of the order of 1.3 kcal mol^{-1} (25 °C, pH = 7, μ = 1 M). The corresponding parameters depend strongly on the composition of solvent, ionic strength and pH. Some parameters characteristic to α-helix-random coil transitions for some proteins (according to [1]) and to nucleic acid double helix-random coil transitions for some types of DNA (according to [47]) are given in Table 5.6. The determination of parameters s and σ was carried out for all the 20 amino acids [48], by measurements either in the corresponding homopolypeptides, or—in the case of hydrophobic amino acids—on soluble homopolypeptides which contain a certain percentage of the amino acid of interest.

The ΔH_c values from Table 5.6 provide a measure of the rigidity of the corresponding helical configurations. The α-helical structures of

TABLE 5.6 **Data for helix-coil transitions**

Polymer	σ	T_{melt} °C	ΔH_c (kcal mol^{-1})	ΔS_c(e.u.)
poly-Gly	0.016	—	—0.49	—3.15
poly-*L*-Ala	0.012	—	—0.19	—0.55
poly-*L*-Leu	0.05—0.011	—	+0.10	+0.8
poly-*L*-Val	0.011	—	+0.21	+1.45
poly-*L*-Glu (neutral)	2.5×10^{-3}	—	—1.0	—2.7
poly-*L*-Lys (neutral)	2.5×10^{-3}	—	—0.8	—2.4
poly-dAT (0% GC)	0.1	—	7.56	—
DNA Cl perfrigens (31%GC)	—	55.0	7.73	23.4
DNA phage T2 (34%GC)	—	75.0	7.8	23.4
DNA huck sperm (41%GC)	—	60.0	7.84	23.6
DNA M. lysodeiktikus (72%GC)	—	79.0	8.52	24.0

proteins should be melted relatively easily and forced into other conformations by interactions with other molecules, while the DNA double helix should be sensibly more rigid, according to this criterion.

5.5.2. *Kinetic aspects of conformational transitions*

Studies on the kinetics of protein and nucleic acid renaturation processes, that is of the transition from random coil to a crystalline structure, have been carried out especially by means of n.m.r., ultrasound dispersion, optical rotation and dielectric methods [48, 49]. These studies have revealed that the kinetics of renaturation process involves some steps of different rates, the slowest being the initial nucleation step. Also, the pathway covered from random coil to the ordered structure does not chaotically go through all imaginable intermediary configurations, but only through a very limited number of configurations, which allow these processes to be much faster.

As regards various elementary processes, the formation and breaking of hydrophobic interactions (10^8—10^9 M^{-1} s^{-1} and 10^2—10^4 s^{-1}, respectively) and of the hydrogen bonds (formation: 5×10^9 M^{-1} s^{-1} for both polynucleotides and polypeptides), they are diffusion-controlled [49, 50]. Other steps—especially the nucleation stages—are by far slower.

The renaturation processes of proteins consist of at least two stages. Thus, the renaturation of staphylococcal nuclease following acidic denaturation, comprises a slow strongly temperature dependent step. Both processes fit a first-order kinetics and pass through an intermediary stage [48c]. The rate constants for these processes would be *ca.* 2 s^{-1} and 12 s^{-1}, respectively [49]. For the coil → α-helix transition of polybenzoylglutamate, the overall process is characterised by an activation enthalpy, $\Delta H^{\neq} = 4$ kcal mol^{-1} and by an activation entropy $\Delta S^{\neq} = -2.6$ e.u. [50a]. For the renaturation of acid-denatured chymotrypsin and trypsin, the kinetics is also of first order, the overall rate constant is 10—100 s^{-1} [49]. Other studies concerning the renaturation of a series of enzymes shows that the presence of substrate, and co-factors increases the rate and degree of renaturation (reactivation). In certain enzymes—e.g. fumarase, enolase, aldolase—the renaturation is fast; in others—e. g. lactate-dehydrogenase—the reactivation is slow: the overall restructuring of the polypeptide chain lasts *ca.* 1 min, other slighter changes, regarding, this time, enzymatic reactivation, occur at a sensitively slower pace [51].

At renaturation of nucleic acids, the nucleation consists of the formation of three base-pairs, in two stages, with the addition of the third pair as a rate-determining step and with high entropy consumption [52]. The rate of such a process (i.e. of first order) of renaturation of poly-(dAT) in aqueous solutions, by pH modification, at 10 °C, is of 0.26 s^{-1} [50b]; for the nucleation of yeast Ala-tRNA, of *ca.* 10^4 s^{-1} (at 15—85 °C) [53]. The growth of double helix, by stacking, is much faster, *ca.* 10^7 s^{-1} per base-pair [49, 53].

REFERENCES

1. A. M. Liquori, *Quart. Rev. Biophysics*, **2,** 65 (1969).
2. M. Sundaralingam, *Biopolymers*, **7,** 821 (1969); (a). E. Soru, *Medical Biochemistry* (in Romanian), Medical Press, Bucharest, (1959), vol. 1, Ch. 1; (b). Cl. Nicolau and Z. Simon, *Molecular Biophysics* (in Romanian), Scientific Publishing House, Bucharest, (1968), Ch. 3.
3. R. Langridge, D. A. Marvin, W. E. Seeds, H. R. Wilson, A. W. Hooper, L. D. Hamilton, M. Spencer and M. H. F. Wilkins, *J. Molec. Biol.*, **2,** 38 (1960); **3,** 547 (1961).
4. H. A. Scheraga, *Chem. Rev.*, **71,** 193 (1971).
5. A. M. Skvortsov, T. M. Birshtein and A. D. Zalevski, *Molek. Biol.*, **5,** 69 (1971)

6(a). G. N. Ramachandran, in: *Symmetry Funct. Biol. Systems Macromolec. Level*, A. Engstrom (editor), W. Almqvist, Stockholm, (1969) p. 79; (b). A. V. Lakshminarayanan and V. Sasisekharan, *Biochim. Biophys. Acta*, **204,** 49 (1970); (c). B. K. Ponuswamy and V. Sasisekharan, *Biophys. Acta*, **221,** 153, 159 (1970).

7. J. L. De Coen, G. Elefante, A. M. Liquori and A. Damiani, *Nature*, **216,** 910 (1967).

8(a). J. C. Kendrew, R. E. Dickerson, B. E. Strandberg, B. G. Hart, D. R. Davies, D. C. Phillips and V. C. Shore, *Nature*, **185,** 422 (1960); (b). J. C. Kendrew, H. C. Watson, B. E. Strandberg, E. R. Dickerson, D. C. Phillips and V. C. Shoer *Nature*, **190,** 666 (1961); (c) H. C. Watson, in *Progress in Stereochemistry*, (1968).

9. M. O. Dayhoff and R. N. Eck (editors), *Atlas of Protein Sequence and Structure*, National Biomedical Research Foundation, (1968).
10. J. M. Klotz, N. R. Langerman and D. W. Darnell, *Ann. Rev. Biochem.*, **39,** 25 (1970).
11. J. W. Prothero, *Biophys. J.*, **6,** 367 (1966).
12. D. Kotelchuck and H. A. Scheraga, *Proc. Nat. Acad. Sci. U.S.A.*, **62,** 14 (1969).
13. O. B. Ptitsyn, *Molec. Biol.*, **3,** 627 (1969); *J. Molec. Biol.*, **42,** 50 (1969).

14. O. B. Ptitsyn and A. V. Finkelshtein, *Biofizika*, **15**, 757 (1970).
15. B. Robson and R. H. Pain, *J. Molec. Biol.*, **58**, 237 (1971).
16. G. M. Lipkind and E. M. Popov, *Molek. Biol.*, **5**, 667 (1971).
17. V. I. Vorobiev and T. M. Birshtein, *Molek. Biol.*, **5**, 327 (1971).
18. T. T. Wu and E. A. Kabat, *Proc. Nat. Acad. Sci. U.S.A.*, **68**, 1501, (1971).
19. Ch. J. Epstein, R. F. Goldberger and Chr. B. Anfinsen, *Cold Spring Harbour Symp. Quant. Biol.*, **28**, 439 (1963).
20. J. A. Rupley, *Struct. Stab. Biol. Macromolecules*, **2**, 291 (1969).
21. M. H. Klapper, *Biochim. Biophys. Acta*, **229**, 557 (1971).
22. O. B. Ptitsyn, *Uspekhi Sovrem. Biol. (Advances of Modern Biol.)*, **69**, 26 (1970).
23. J. L. De Coen, *J. Molec. Biol.*, **49**, 405 (1970).
24. A. Ikai and C. Tanford, *Nature*, **230**, 100 (1971).
25. P. N. Lewis, F. A. Momany and H. A. Scheraga, *Proc. Nat. Acad. U.S.A.*, **68**, 2293 (1971).
26. H. D. Welscher, *Int. J. Protein Res.*, **1**, 243 (1969).
27a. I. M. Klotz, *Arch. Biochem. Biophys.*, **138**, 704, (1970) 27b. B. H. Nicholson, *Biochem. J.*, **123**, 117 (1971).
28. A. A. Zamiatnin, *Biofizika*, **16**, 163 (1971).
29. C. Makino *et al.*, *Biopolymers*, **6**, 551 (1970); H. Ingwal *et al.*, *idem*, **331**; cited in *Nature*, **218**, 917 (1968).
30. O. B. Ptitsyn, *Usp. Sovr. Biol. (Advan. Med. Biol.)*, **63**, 3 (1967).
31. V. A. Pehelin, A. V. Kolynskaia and V. N. Izmailova, *Dokl. Ak. Nauk SSSR (Proc. U.S.S.R. Acad. Sci.)*, **186**, 139 (1969).
32. (a) L. G. Seldykh, N. V. Seldykh, "Biofizika", **15**, 769 (1970); (b) B. P. Atanasov, *Molek. Biol.*, **4**, 348 (1970).
33. (a) H. P. Avey, M. P. Boles, C. H. Carlisle, S. A. Evans, S. J. Morris, P. A. Palmer, B. A. Woolhouse and S. Shall, *Nature*, **213**, 557 (1967); (b) G. Kartha, J. Bello, D. Harker, *ibid.* 862; (c) Cf. *ibid.*, 960.
34. (a) G. M. Edelman, B. A. Cunningham, W. Einar, Gall, P. D. Gotlieb, V. Rutishauer and M. J. Waxdal, *Proc. Nat. Ac. Sci. U.S.A.*, **63**, 78 (1969); (b) S.J. Singer and R. F. Dootittle, *Science*, **153**, 13 (1966); (c) M. J. Poljack, J. Arnzel, H. P. Avey and W. Beck, *Nature, New Biol.*, **235**, 137 (1972); (d) T. Sarma, D. Silverton, D. R. Davies and C. Terry, *J. Biol. Chem.*, **246**, 3753 (1971); cited in *Nature*, **235**, 249 (1972); (e) V. Gheţie, in: *Recent Advan. in Immunology* (in Romanian), N. Nestorescu (editor), Medical Press, Bucharest, (1969).
35. J. Monod, J. Wyman and J. P. Changeux, in: *Theoretical Physics and Biology*, M. Marois (editor), North Holland Publ., Co., Amsterdam, (1969).
36. I. M. Klotz and D. W. Darnell, *Science*, **166**, 126 (1969).
37. B. H. Zimm and N. R. Kallenbach, *Ann. Rev. Phys. Chem.*, **13**, 171 (1962) and works cited therein.
38. A. M. Lesk, *J. Theoret. Biol.*, **22**, 537 (1969).
39. (a) H. G. Zachau, *Angew. Chemie*, **81**, 645 (1969); (b) P. G. Conors, M. Labanauskas and W. W. Bekman, *Science*, **166**, 1528 (1969); (c) G. R. Phillips, *Nature*, **223**, 374 (1969).
40. Bram—cited in *Nature*, **232**, 374 (1971).

41. Cited in *Nature*, **234,** 128 (1970); (a) A. Achter, G. Felsenfeld, *Biopolymers*, **10,** 1625 (1971); (b) B. Formoso, J. Tinoco jr., *idem*, 1533; (c) D. R. Davies, J. N. Davidson, *idem*, 1455; (d) C. Maurizot, M. Blicharki, J. Brahms, *ibid*. 1429.
42. I. Tinoco, jr., O. C. Uhlenbeck and M. D. Levine, *Nature*, **230,** 362 (1971).
43. T. M. Birshtein and O. B. Ptitsyn, *Conformation of Macromolecules* (in Russian), Publishing House for Science, Moscow, (1964), Chs. 8 and 9.
44. B. H. Zimm and J. K. Bragg, *J. Chem. Phys.*, **31,** 526 (1959); S. Lifson and A. Roig, *idem*, **34,** 1963 (1961).
45. (a) G. I. Lichtenstein and T. V. Troshkina, *Molek. Biol.*, **2,** 654, (1968); (b) T. M. Birshtein, A. M. Eliashevich and A. M. Skvortzov, *Idem*, **5,** 58 (1971).
46. S. Lewin, *J. Theoret. Biol.*, **17**, 18 (1967).
47. H. Klump and T. Ackermann, *Biopolymers*, **10,** 513 (1971).
48. (a) C. Van Dreele, *Macromolecules*, **4,** 396, 408 (1971); G. Ananthanarayan, *ibid.*, 417; (b) T. W. Epstein, *J. Molec. Biol.*, **60,** 499 (1971); *cited* in *Nature*, **233,** 523 (1971).
49. A. N. Schechter, *Science*, **170,** 273 (1970).
50. (a) M. Davies, *Chem. Ind.*, **35,** 1191 (1968); (b) T. M. Hicky and E. Hamori, *J. Molec. Biol.*, **57,** 359 (1971).
51. J. W. Teipel and D. E. Koshland jr., *Biochemistry*, **10,** 792, 798, 805 (1971).
52. J. M. Craig, M. Crothers, R. Doty, *J. Molec. Biol.*, **62,** 383 (1971); cited in *Nature*, **235,** 77 (1972).
53. D. Riesner, R. Rocmer and G. Maass, *Eur. J. Biochem.*, **15,** 85 (1970).
54. V. I. Lim and O. B. Ptitsyn, *Biofizika*, **17,** 21 (1972).
55. A. V. Finkelshtein and O. B. Ptitsyn, *J. Molec., Biol.* **62,** 613 (1971).
56. Ch. Delisi and M. Crothers, *Proc. Nat. Ac. Sci. U.S.A.*, **68,** 2682, (1971).
57. Comment, *Nature*, **242,** 555 (1973).
58. J. L. Crawford, W. N. Lypscomb and Ch. G. Schellman, *Proc. Nat. Ac. Sci. U.S.A.*, **70,** 538 (1973).
59. P. Y. Chou and G. D. Fasman, *J. Molec. Biol.*, **74,** 263 (1973).
60. P. B. Ptitsyn, *Pure and Appl. Chem.*, **31,** 227 (1972).
61. (a) K. Nagano, *J. Molec. Biol.*, **75,** 401 (1973); (b) P. K. Ponnuswany, P. K. Warme and H. A. Scherega, *Proc. Nat. Ac. Sci. U.S.A.*, **70,** 830 (1973).
62. (a) J. Mc. Alister, N. Iathindra and M. Sundaralingam, *Biochemistry*, **12,** 1189 (1973); (b) L. B. Kier, *J. Theoret. Biol.*, **40,** 211 (1973); (c) H. J. R. Weintraub and A. J. Hopfinger, *J. Theoret. Biol.*, **41,** 53 (1973).
63. O. B. Ptitsyn, *Vestnik Akad. Nauk S.S.S.R.* 1973 (5) 57.
64. (a) V. I. Lim, *Dokl. Akad. Nauk S.S.S.R.*, **203,** 480 (1972); (b) A. V. Finkelshtein, *ibid.* **207,** 1486 (1973); (c) O. B. Ptitsyn, *Report 6th Jena Molecular Biology Symposium*, (1973).
65. R. Masson, A. Wutrich, *FEBS Latt*, **31,** 114 (1973); H. Karplus, A. B. Snyder, B. Sykes, *Biochemistry*, **12,** 1323 (1973); cited in *Nature*, **243,** 57 (1973).
66. P.-H. von Hippel and J. D. McGhee, *Ann. Rev. Biochem.*, **41,** 232, (1972).
67. (a) W. Min Jou, G. Haegeman, M. Isobaert and W. Fiers, *Nature*, **237,** 82 (1972); (b) L. A. Ball, *J. Theoret. Biol.*, **41,** 243 (1973); (c) R. Figueroa, A. Soto, G. Gonzalez, M. Piebes, C. Romero and J. C. Toha, *ibid.* **36,** 321 (1972).
68. (a) O. C. Uhlenbeck, Ph. D. Borer, B. Denyler and J. Tinoco jr., *J. Molec. Biol.*, **73,** 483, (1973); (b) J. Gralle and D. M. Crothers, *ibid.* **78,** 301 (1973).

6. Specific interactions

Throughout various biological processes, as for example enzymic reactions, antigen-antibody reactions, inhibition of a certain operator of the genome by a repressor, a molecular species A_0 specifically recognises a certain molecular species B_0, and only that one, out of a great number N of existing B_i species. Thus, many enzymes catalyse the reaction of a single substrate species (here B_0) out of several similar substrate species within the same organism. A certain antibody (here A_0) recognises only a single foreign antigen but not the very large number ($N \cong 10^5$ or may be more) of self-antigenic determinants (here B_i, $i \neq 0$). A certain repressor recognises a certain operator gene and does not bind to other nucleotidic sequences within the respective genome ($N \cong 10^6$ for the genome of bacterias like *Escherichia coli*).

Since the time of Paul Ehrlich [1], the assumption is implicit that recognition implies the formation of a complex between A_0, termed here the recogniser and the correct partner B_0, i.e. the affinity of A_0 towards B_0 is higher than for the "false" partners B_i ($i \neq 0$). A high specificity of such a recognition process seems to be determined by a high molecular size of the A and B partners. A micromolecular ion, S^{2-} for example, precipitates as sulphides, tens of cations out of some hundreds of cations in analytic chemistry; ethylenediamine tetra-acetic acid with a higher molecular weight is more selective—it complexes only, perhaps, 10 or 20 cations. All molecular species able to give highly specific recognitions—antibodies, repressors, enzymes, are macromolecules and even the common antibiotics are quite large molecules. The relation of specificity with molecular size was discussed in relation to DNA-RNA hybridisation competition experiments by Thomas [2a] and by Conaughy and McCarthy [2b]. The precision of DNA replication and transcription and the probability of the incorporation of a false base were related to affinity differencies (free enthalpies of binding) by Volkenshtein and Eliashevich [3]. A general discussion of recognition processes implying biopolymers was given by Bradley [4].

Our aim is to give a quantitative approximate treatment of the specificity - affinity - molecular size relations based on some thermodynamic and probabilistic principles [5] and on some considerations concerning spatial conformation and intermolecular forces.

6.1. **Quantitative characterisation of interaction specificity**

The problem of specific interaction, or of recongnition, may be defined as follows: a molecular species A_0 (the recogniser) must be able to discern a certain molecular species B_0, the "correct effector" out of N such species $B_0, B_1, ..., B_N$ of "correct" and "false" effectors. Also, the probability of all false recognitions should be lower than a threshold p_f. The degree of specificity for this process is expressed by the number N of effectors among which the recogniser must discern the correct one and by the threshold p_f which indicates the precision of the process. The recognition prɔcesess must be at first formulated in molecular terms, the pertinent structural characteristics of the A and B species must be specified and then relations between the degree of specificity, structural characteristics and molecular size (more exactly of combining sites on A and B) should be inferred.

The number N of effectors comprises only those false molecular species whose "recognition" is harmful to the organism. Thus in antibody elicitation by a foreign antigenic determinant, N contains only those self-antigenic determinants which may come into contact with antibodies, and not those which never come into contact with immunocytes (possibly some intracellular proteins, the proteins of the crystalline lens, *etc.* [6]). Or else, in the process of transport RNA amino-acylation, each amino acid must be recognised by a certain amino-acylating enzyme out of a total of 20—40 such enzymes within the organism; complex formations with other enzymes does not interfere with the precision of tRNA amino-acylation.

We shall assume that the decisive step in recognition, at the molecular scale, is the formation of an A_0B_0 complex:

$$A_0 + B_i \rightarrow A_0B_i \begin{cases} \text{yes, for } i = 0 \\ \text{no, for } i \neq 0 \end{cases} \tag{6.1}$$

We shall also consider a state of thermodynamic equilibrium and a 1 : 1 stoichiometry in the formation of these complexes. The thermodynamic equilibrium hypothesis stands if only intermolecular forces and not valence ones intervene in complex formation; this is due to high reaction rates of complex formation (see section 5.5.2). More general stoichiometries for antigen-antibody interactions were discussed by Palmiter and Aladjem [7a] and a kinetic treatment for precipitation was given by Nielsen [7b]. Including a more general stoichiometry and non-equilibrium states in our discussion of complex formation is not likely to bear much on the results, as the same factors favour or disfavour complex formation or dissociation, irrespective of stoichiometry and departure from equilibrium. For "kinetic" recognitions—for example enzymic reactions—we may adopt the viewpoint of the absolute reaction-rate theory, considering the A_0 and B_i species as reactants and, as activated complexes, the A_0B_i complex (for nuclear configurations corresponding to the activated complex).

If complex formation is a simple association-dissociation equilibrium, for each process:

$$A_0 + B_i \rightleftarrows A_0B_i \therefore \frac{[(A_0B_i)]}{[(A_0][B_i]} = K_{0i} = e^{-\frac{G_{0i}}{RT}}; \quad i = 0, 1, \ldots, N \quad (6.2)$$

while the threshold p_f of all "false" recongnitions will be given by the relative concentration of all false A_0B_i complexes [4a]:

$$p_f = \frac{\sum_{i=1}^{N} [(A_0B_i)]}{[(A_0B_0)] + \sum_{i=1}^{N} [(A_0B_i)]} = \frac{\sum_{i=1}^{N} [B_i] \cdot K_{0i}}{[B_0] \cdot K_{00} + \sum_{i=1}^{N} [B_i] \cdot K_{0i}} \quad (6.3)$$

where K_{0i} are the association constants; G_{0i}, the affinity of A_0 for B_i; $[B_0]$, $[B_i]$, *etc.* the respective concentrations. The last relation (6.3) is inferred for a low degree of B_0, B_i, ..., B_N saturation with A_0. For large A_0 quantities, such that all B_i's are saturated with A_0, $[A_0B_i]$ is practically equal to $[B_i]$ and the recognition process breaks down [19].

If the A_0—B_i interactions have a co-operative character, at the limit, recognition will represent precipitation of A_0 only by B_0 and not also by the false B_i effectors. The following solubility product conditions will have to be satisfied in this case [5d]:

$$[A_0] \cdot [B_0] \geqslant K_{00}^{-1}; \quad [A_0] \cdot [B_i] < K_{0i}^{-1} \text{ for } i = 1, 2, \ldots, N. \quad (6.4)$$

If co-precipitation is neglected, $p_f = 0$. Equations (6.4) are also valid for a large excess of B_0, B_i, ... over A_0. Here also, if some $[A_0]$ $[B_i]$ products are higher than the corresponding solubility products and there is sufficient time for different A_0B_i precipitates to equilibrate A_0 between them, the process may be governed by an equation of type (6.3) with a finite error probability. Excess of A_0 over the B_i values in this latter case may also drastically reduce the precision of recognition.

According to Riggs, Suzuki and Bourgeois [8] the *E. coli* lac repressor-operator interaction is a simple association-dissociation equilibrium. Antigen-antibody neutralisation and possibly also the elicitation of antibody response are precipitation reactions, although not always of a 1 : 1 stoichiometry [7a, 9].

With these assumptions concerning complex formation and thermodynamic equilibrium it is clear that the recogniser A_0 must have the highest affinity—G_{00} for the "correct" partner B_0. The structural differences between B_0 and the false B_i values will lower the corresponding—G_{0i} affinities with

$$\Delta G_i \equiv G_{0i} - G_{00}. \tag{6.5}$$

Thus, a highly specific recognition requires sufficiently high ΔG_i differences between the affinity of A_0 towards the correct B_0-effector and the false ones, as well as a sufficient excess of B_0 over A_0. The "recogniser" A_0 must be able to discern with high enough ΔG_i-differences a certain B_0 out of a total of N different B_i-effectors. Small structural differrences on the B_i regions which come into contact with A_0 (combining sites) will give in most cases small ΔG_i differences. Therefore the combining sites on A_0 and B_i must be large enough to allow sufficiently large ΔG_i-differences for all the N (presumablly different) combining sites on the "false" B_i effectors. One can understand why a high specificity—large N, small p_f, large ΔG_i values (required eventually as a result of high B_i, low B_0 concentrations)—require large combining sites and a rather rigid combining site on the recogniser A_0 (in order to ensure a steric complementarity only against the B_0 combining site). This rigidity of the recogniser combining site requires a supplementary part of the A_0 molecule to ensure the rigidisation. Thus the minimal molecular size will be given by the "recogniser site" which comprises also those parts of the A_0 and B_i molecules where structural modifications produce variations in the shape of the combining sites.

6.2. The role of steric configuration

Steric configuration is recognised as the most important parameter in determining the affinity of a molecule towards the sites determining various biological activities — the "lock and key" theory [1]. Unfortunately this parameter is also the most difficult to characterise quantitatively if we do not want to resort to a hopelessly large number of interatomic bond lengths and angles and van der Waals radii to describe it.

The importance of steric configuration over that of chemical groups results, for example, from the fact that enzymes transform several similar substrates of the same chirality (although with different rates) but usually not at all the optical antipode of even the most active substrate. In antigen-antibody recognitions, antigenic determinants in polypeptides have a "conformational" rather than a sequential character [10]. Thus the α-helical poly-(Tyr-Ala-Glu) peptide elicits an antibody which gives no cross reaction with the non-helical Tyr-Ala-Glu tripeptide acylated on the ε-amino groups of polylysine; the reciprocal non-cross-reactivity is also demonstrated. There are also examples when substitution of a charged $N(CH_3)_3^+$ group by the $C(CH_3)_3$ group (which has the same shape and volume) in an antigenic determinant, does not lower sensibly the affinity of the antibody towards the determinant [11]. Nevertheless, a structural modification which does not affect shape and volume but changes intermolecular force characteristics, usually produces also an affinity decrease —*ca.* 1.7 kcal mol^{-1} for substitutions of $-CO_2^-$ by $-NO_2$ (section 3.3.4).

The first attempt to correlate the decrease of affinity (more exactly the decrease of biological activity) with structural modifications including also various types of steric changes seems to be the one of Sneath [12]. He correlates decrease in biological activity in substituted oligopeptides (as compared to the hormonal oligopeptide as standard) with the structural modifications produced by the respective substitutions. These modifications refer to the presence or absence of 134 characters, as pyrrolic N⟨, HN⟨ or H_2N- groups, ramifications in the α, β or γ positions of the main amino acidic side chains, inertial moments within certain limits for rotations about certain types of bonds, etc. Possibly most characters do not bear on the decrease of affinity and therefore the linear correlation coefficients are low, $-r = 0.389$; 0.135 and 0.750 for 29 oxytocytic, 24 pressoric and 12 hypertensinic derivatives, respectively. (See Appendix 2 for a calculation of r).

A quantitative characterisation of steric dissimilarities between two molecules could be obtained by the procedure of Rao and Rossmann [19], employed by them in comparing suprasecondary structures in proteins. They superposed the two similar structures by a geometric technique so as to minimise the sum of quadratic distances between all atoms assumed to be equivalent. This sum is the measure of steric dissimilarity. Computing facilities are required in this procedure.

Taft's steric constants E_s represents a general (screening) effect on a reaction centre, in a micromolecular reaction, of various subtituents [25]. One could hardly expect, therefore, to correlate these steric constants with the effects of steric fit. Really, an attempt to correlate α-chymotrypsin catalysed hydrolysis rates of acyl substituted *p*-nitrophenyl esters with σ, π and E_s constants succeeded only for a series of nine compounds out of the 42 esters studied [26].

6.2.1. *Minimal steric difference (MSD) parameter*

This was introduced by Simon and Szabaday [20]. The steric configuration of sites responsible for biological activities is, usually, only guessed from the configuration of the natural substrate molecule, i.e. the molecule with the (assumed) maximum effect with respect to the given biological activity. This natural substrate molecule may possess several approximately isoenergetic spatial configurations and we do not know which one "enters" the site of biological action.

The minimal steric difference (MSD) between a given molecule and the respective natural substrate is the non-overlapping volume of those lowest energy configurations which allow a maximum of spatial overlap between the two molecules. For the sake of simplicity one attributes to the volumes of all non-overlapping groups, XH_n ($n = 0, 1, ...$) the value 1 if X is a second period element, 1.5 if X belongs to the third period and 2 for lower periods.

In Table 6.1 the MSD values are correlated with the maximum hydrolysis rates (at substrate saturation) of the methyl esters of *N*-acetylamino acids (as listed by Knowles [21]: $CH_3CONHCHRCOOCH_3$). In this example, the changing R group is not directly involved in the reaction; the reactivity of the hydrolysed group may depend strongly on parameters such as net atomic charges which are practically unaltered here. Table 6.1 contains the hydrolysis rates $V_{mx}(M^{-1})$ at low substrate

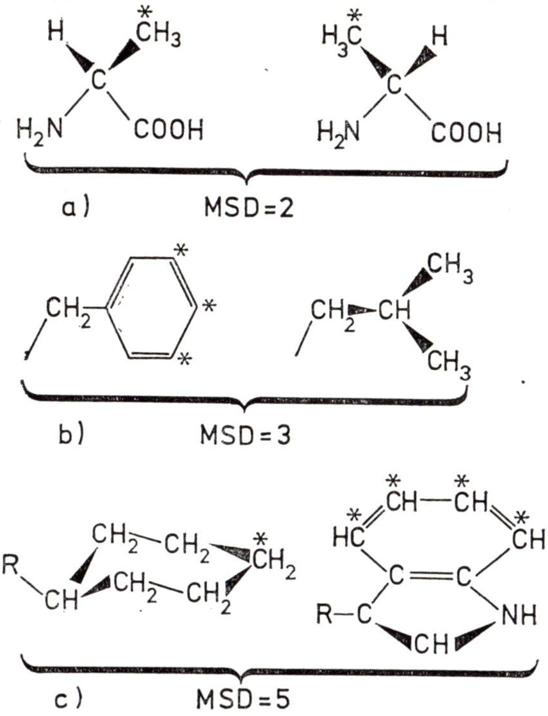

Figure **6.1.** Minimal steric differences. Unsuperposable atoms (groups) are marked by an asterisk (*).
a) D-alanine *vs* L-alanine; b) benzyl *vs* isobutyl; c) R-cyclohexyl *vs* R-indolyl.

a) MSD=2

b) MSD=3

c) MSD=5

TABLE 6.1 **Kinetic parameters for α-chymotripsin hydrolysis**

Amino acid in $CH_2CONHCHRCOOCH_3$	V_{mx}	MSD	HP
Trp	4.2×10^5	0	3.00
Tyr	3.65×10^5	5	2.97
Cyclohexylalanine	8×10^4	5	4.0
Phe	4.2×10^4	4	2.65
α-Aminoheptanoic acid	8.17×10^3	5	3.1
Met	2.3×10^3	5	1.3
α-Aminocaprilic acid	2.12×10^3	4	3.8
Leu	1.59×10^3	6	2.49
Nor-Leu	1.25×10^3	6	2.5
Nor-Val	2.65×10^2	7	1.7
α-Aminobutiric acid	19.6	8	1.1
Ala	1.72	9	0.73
Val	1.35	9	1.69
Gly	9.8×10^{-3}	10	0

concentration together with the MSD values with respect to the triptophane derivative (the molecule with the highest hydrolysis rate in this series) and the Tanford hydrophobicity indexes, HP (kcal mol^{-1}; or interpolated according to them — see Table 3.11, section 3.3.7) of the R side chain.

The log V_{mx} *versus* HP correlation yields a linear correlation coefficient $r = 0.80$ while the log V_{mx} *versus* MSD correlation yields $r = —0.91$ and explains also the lowering of hydrolysis rates for cyclohexylalanine, α-aminoheptanoic and α-amino caprilic acid derivatives.

In other cases the MSD-activity correlation is less satisfactory. In a correlation of MSD with affinity decrease for a series of 92 substituted oligopeptides (see section 6.3) — $r = 0.55$ is obtained [20]; with six parameters characterising the differences in types of intermolecular forces —$r = 0.75$ is obtained [15]. An attempt to correlate MSD values for the purinic and pyrimidinic bases (compared to adenine) with the relative hydrolysis rates of the corresponding ribosides gives a less satisfactory result than the correlation with the π-electronic charge at the glycosidic *N*-atom (—$r = 0.57$, Table 4.3, section 4.3.2). Also in a correlation of Michaelis constant for ATP-phosphorylation by adenosine kinase of twelve different β-ribosides only —$r = 0.63$ is obtained (Table 4.5, section 4.3.2) and even less, —$r = 0.15$, when modifications in the ribose moiety of the nucleosides are also taken into account [20].

Our MSD-procedure has some similarities with the methods of comparing the molecular shape, given by Amoore *et al.* [27] and by Allinger [28]. These methods are much more laborious, require computerisation and do not seem to give better correlations ($r = 0.45$—0.75 for correlations with odour characteristics [27]).

6.2.2. *Correlations with secondary structure parameters*

We tried to correlate the affinity decrease induced by amino-acidic substitutions with parameters for the tendency of the respective amino acids to favour or disfavour the α-helical structure or the β-structure [13]. Some results are listed in Table 6.2. The parameters for α-helical structure used in calculation of Δ(αHe) differences produced by the substitutions are [13 b]: +2 for Ala, Val, Leu and Glu; +1 for Cys, Met, Pro, Phe, Trp, GlN and Arg; 0 for Gly and —1 for Tyr, Ser, Thr, AsN, Asp, Lys, His and the amino acid residue preceding proline. Δβ refers to β-structure—

TABLE 6.2 **Correlation of affinity decrease to changes produced by amino-acidic substitutions in bradykinine**

Substitution	A_i^{exp}					Δ(αHe)	Δβ
	A—BK—TC—PL		a—BK—TC—OA		a-Kin		
	BK	Ac—BK	T24	T36			
Bradykinine	0	0	0	0	0	0	0
Arg/Ala⁹-BK	—	—0.20	—0.52	—0.62	—2.95	1	75
Phe/Ala⁸-BK	—	—0.35	—2.85	—	—2.19	1	—25
Pro/Ala⁷-BK	—1.00	—0.50	—1.61	—	—0.76	1	75
Ser/Ala⁶-BK	—0.38	—0.80	—0.75	—1.62	—1.79	0	50
Phe/Ala⁵-BK	—1.57	—1.35	—3.77	—3.94	—1.93	1	—25
Gly/Ala⁴-BK	—1.58	—1.85	—1.05	—1.42	—0.56	2	25
Pro/Ala³-BK	—2.58	—1.96	—0.76	—1.13	—0.08	3	75
Pro/Ala²-BK	—	—2.63	—0.11	—0.83	—0.15	2	75

favouring character of amino acids (see section 5.2.1). The substances used in correlations were first of all the above-cited three series of oxytocytic, pressoric and hypertensinic derivatives of Sneath [12], whose standard oligopeptides are listed in Figure 6.2. The other oligopeptide series

Oxytocyne:

S—S (bridge between the two Cys)
HCys—Tyr—Ile—GlN—AsN—Cys—Pro—Leu—GlyOH
Activated
standard.
pressors:

S—S (bridge between the two Cys)
HCys—Tyr—Phe—GlN—AsN—Cys—Pro—Arg—GlyOH
Hypertensine: HAsN—Arg—Val—Tyr—Val—His—Pro—PheOH
Bradykinine: HArg—Pro—Pro—Gly—Phe—Ser—Pro—Phe—ArgOH
The way of brady-
quinine bonding to support proteine (*Carrier*):

CH_3
Carrier —NH—CO—NH—(benzene ring)—NH—CO—NH—
Bradykinine—OH

Figure 6.2. Primary structure of some standard oligopeptides.

used for these correlations and also for affinity decrease-intermolecular force correlations are: three tetra-alanine substitution derivatives [14 a]; two derivatives of the Val-(ε-dinitrophenyl-Lys)-Leu-Phe-OC_2H_5 peptide [14b); eight derivatives of an encephalomielitogenic undecapeptide [14c]; three peptidoglican derivatives [14d]; a series of bradykinine derivatives measured in their relative affinity towards antibodies elicited in various rabbits by bradykinine coupled to polylysine (a—BK—TC—PL), ovalbumine (a—BK—TC—OA; T24 and T26) and by akininogen (a—kin) [14e]; a series of bradykinine derivatives (Ac—BK) whose affinities are compared to tritiated acetylbradykinine [14f]. Figure 6.2 gives also the primary structure of bradykinine; H means the amino terminal, OH the carbonyl terminal. Details concerning the "biological activity" and affinity difference calculations are listed in Appendix 3.

As to results, the affinity decreases *versus* Δ(αHe) correlations for the three series of Sneath are, respectively, $-r = 0.537$, 0.19 and 0.0. A good correlation with Δ(αHe), $r = -0.95$ is obtained only for the relative affinity of bradykinine towards antibodies elicited by bradykinine coupled to polylysine (Table 6.2); if the antibodies are elicited by bradykinine coupled to ovalbumine or by akininogen the correlation is practically absent ($r \cong 0$) [13].

The very poor results of these attempts to correlate the decrease of affinity to secondary structural parameters for amino acids indicates that the steric configuration is determined by interactions between various amino acidic residues. The single satisfactory correlation is obtained for the antibody elicited by the bradykine—two positive charges at pH = 7, coupled to polylysine—highly positively charged at pH = 7. Bradykinine does not interact with the carrier in this case and its steric configuration—as antigenic determinant—is the same as for free bradykinine. Ovalbumine-carried bradykinine may have a quite different steric configuration from free bradykinine, due to interactions with the carrier protein, ovalbumine, and the same may be true for the bradykinine residue in akininogen [13b].

The global steric configuration of a peptidic chain is determined by interactions, *via* intermolecular forces, between non-neighbouring amino-acidic residues too. Thus, the reasonable correlation, described in the following paragraph (section 6.3), of affinity decrease with changes in intermolecular force parameters does not contradict Sela's statement [10] that antigenic determinants are of a fairly conformational nature.

6.2.3. *Some conclusions*

The rather scarce success of correlations of biological activity *versus* MSD indicates that only the steric fit complementarity does not suffice for a high affinity of interaction; the steric fit should juxtapose also, on the two partners, atoms complementary with respect to intermolecular forces. These very strict requirements are true if the combining sites on both the recogniser A_0 and the effector B_0 are quite rigid or are in a cavity-excrescence relation. Steric requirements may be lowered if these combining sites are non-rigid and situated on the surface of the molecules. Thus, the affinity of the antibody elicited against *p*-(*p'*-azophenylazo)-benzoate coupled to a protein towards substituted benzoate ions presents the following features. A stringent steric fit is required for the first benzenic ring (the one in the benzoate moiety); affinity increases of *ca.* 1.0 kcal mol^{-1} are obtained for *para*-substitutions. Much looser steric requirements are imposed for the second benzenic ring and for the second azo-group ([11] pp. 56—61). It looks like only the *p*-azobenzoate moiety of the antigen enters a "lock" in the elicited antibody. Also, if the oxytocytic activity decrease of oxytocyne derivatives obtained by amino-acidic substitution in position 3, 4 and 8 only (Table 6.4) is correlated with the decrease calculated by equation (6.7) for changes in intermolecular force parameters, a very good correlation is obtained ($r = 0.95$). The correlation becomes much worse if oxytocyne substitution derivatives in positions 2 and 5 are also considered ($r = 0.74$). It seems that only the side chains of amino-acidic residues in positions 2 and 5 of oxytocyne enter some cavities of the biological site of action, the other side chains may interact with groups on the surface of the site of action [20].

6.3. The rôle of intermolecular forces

In the previous paragraph we emphasised the effect of steric configuration upon the A—B affinity in specific interaction. A "key into lock" fit of B_0 and A_0 partners is certainly necessary for a high affinity, but it is not sufficient; for example a steric fit between two regions, both charged positively or negatively, will add a large electrostatic repulsion term to $-G_{0i}$. The same would be true for a steric fit between a hydrophobic and a hydrogen bond forming region. The importance of intermolecular forces results also from the rule that a substance dissolves into liquids of similar structure.

TABLE 6.3 **Some A_i^{exp} and σ_{ji} values for various amino-acidic substitutions** (from [15], Table 3)

Amino-acidic substitution	*Recorded A_i^{exp} values*	σ_{1i}	σ_{2i}	σ_{3i}	σ_{4i}	σ_{5i}	σ_{6i}	MSD
Ala/Gly	1.58; 1.05; 1.42 0.00; 0.30; 0.89 0.56; 1.51; 0.13	0.73	0	0	0	0	0	1
Ala/Phe	2.85; 2.19; 1.57 3.94; 3.77; 1.93 0.05; 0.30; 1.37	1.92	1.5	0	0	0	0	6
Phe/Tyr	1.15; 1.00	0.32	0	0	0	1	1	1
Leu/Lys	0.76	0.9	0	0	1	1.25	0	3
Phe/Tyr+Lys/Leu	2.65	1.22	0	0	1	2.25	1	4
Trp/Ileu	4.05	0	2.5	1	0	1	1	8
Ala/Ser	0.38; 0.75; 1.62 1.79; 1.5	0.69	0	0	0	1.5	0	1

In order to test the importance of intermolecular forces for specific interactions, large groups of substances must be taken into account and statistics applied. We will use the linear correlation coefficient r (see Appendix 2) for this purpose, as in the preceding paragraph. Correlation of affinity decreases (towards the site of biological action) produced by amino-acidic substitutions in a standard oligopeptide to changes induced in types of intermolecular forces (characterised in a binary system according to Table 3.14, section 3.4) was carried out in [13a]. The relative biological activities, log A_i, are those discussed also in section 6.2. We have also considered the relative affinity, for the antibody elicited by the standard, of some bradykinine derivatives obtained by nitration or acetylation at amino-acidic side chains and of some di- to octa-peptides derived from bradykinine by the cut-off of 1 to 7 amino acids from either the amino or the carboxyl terminal [14e]. The change in type of intermolecular force was considered as the number of forces which appear and disappear on substitutions which relate the oligopeptide to the standard [13]. The correlation coefficients thus obtained are usually in the range $-r = 0.5 \ldots 0.8$. The mean affinity decrease per type of character for the studied series is 0.9 kcal mol^{-1} [13a]. For the 190 possible substitutions between the 20 natural amino acids, there is a mean change of 2.5 characters or 2.2 kcal mol^{-1} affinity decrease per substitution. A direct comparison of affinity decreases for a total of 1, 2 and 3 amino-acidic substitutions gives a mean affinity decrease of 1.7 kcal mol^{-1}. Thus 2.0 kcal mol^{-1} may be taken as a mean affinity decrease per amino-acidic substitution [5d].

A more quantitative evaluation of the mean effect of different types of intermolecular forces upon interaction specificity was performed by a multiple correlation method [15]. This method considers an invariant "molecular core", with m "positions" in which structural changes, characterised by σ_{ji} parameters occur. In our case the σ_{ji} parameters are a measure for the change in "jth" type of intermolecular force in the compound "i", as compared to the standard compound. If A_i^{exp} is the "activity" of compound i as compared to the standard (see Appendix 3) an optimum linear equation:

$$\log A_i^{calc} = \alpha + \sum_j \beta_j \sigma_{ji}; \quad j = 1, 2 \ldots m; \quad i = 1, 2 \ldots N \qquad (6.6)$$

is set up. The α and β_j parameters are determined by minimisation of the sum of quadratic $A_i^{calc} - A_i^{exp}$ differences (Appendix 4). The β_j parameters are a measure of the mean weight of jth type of intermolecular force.

The studied correlation [15] refers to the relative hormonal effect, or the relative affinity to an antibody of oligopeptides considered a standard one. Of the series mentioned before, all except the last four have been used together with eight relative meningogenic activities of amino acid substitution derivatives in a encephalomyelitogenic undecapeptide [14c] and three relative affinities of tetra-alanine derivatives for the antibody elicited against tetra-alanine [14a]. The amino-acidic substitutions and the A_i^{exp} values for some of the $N = 92$ derivatives considered are listed in Table 6.3. The types of intermolecular forces $\sigma_{i1} - \sigma_{i6}$ refer to those in Table 3.14, section 3.4. σ_{i1} pertain to the hydrophobic increment, HP of the amino-acidic side chains (G_x of Table 3.11, section 3.3.7). σ_{2i} is the aromatic character increment (AR, section 3.3.8) and its value is 1.5 kcal mol^{-1} for Phe, Tyr, 1.0 kcal mol^{-1} for $HisH^+$ ($pH = 7$) and 2.5 kcal mol^{-1} for Trp. σ_{3i} refers to the presence or absence of electron-donor

TABLE 6.4 **Activity decrease-intermolecular force correlations for amino acid substitution derivatives of oexytocyne**

No.	*Substitution*	A_i^{exp}	A_i^{calc} equation (6.7)	*No.*	*Substitution* equation (6.7)	A_i^{exp}	A_i^{calc}
1	Phe_2	1.15	1.61	22	Phe_2Arg_8	2.35	2.25
2	Ser_2	5.0	2.94	23	Ser_2 His_3	5.0	4.29
3	Phe_3	1.25	1.51	24	Ser_2Lys_8	5.0	3.49
4	Tyr_3	3.65	2.59	25	His_2Phe_3	5.0	2.93
5	Trp_3	4.05	4.37	26	Phe_3Lys_8	1.95	2.06
6	Leu_3	1.0	0.52	27	Phe_3Arg_8	0.8	2.15
7	Val_3	0.9	0.60	28	Phe_3His_8	2.4	2.79
8	Ser_4	0.3	0.64	29	Tr_3Lys_8	4.6	3 14
9	Ala_4	1.1	1.25	30	Trp_3Lys_8	>5.0	3.92
10	AsN_4	0.6	0.46	31	Ser_4Ile_8	0.5	0.71
11	Ser_5	2.8	0.65	32	AsN_4GlN_5	3.05	0.47
12	Ala_5	>5.0	1.24	33	$Phe_2Phe_3Lys_8$	3.2	3.22
13	GlN_5	2.65	0.46	34	$Phe_2Phe_3Arg_8$	3.35	3.31
14	Val_5	5.0	1.39	35	$Phe_2Tyr_3Lys_8$	>5.0	4.30
15	Ile_8	0.2	0.52	36	$Ser_2His_3Lys_8$	>5.0	4.84
16	Val_8	0.35	0.54	37	$His_2Ser_3Lys_8$	>5.0	3.30
17	Lys_8	0.75	1.00	38	$His_2Phe_3Lys_8$	>5.0	3.49
18	Arg_8	0.75	1.09	39	$Phe_3AsN_4Lys_8$	2.2	2.07
19	Phe_2Phe_3	2.1	2.67	40	$Phe_3Ala_4Lys_8$	2.95	2.86
20	Phe_2Tyr_3	>5.0	3.75	41	$Phe_3Ser_4Lys_8$	2.7	2.25
21	Phe_2Lys_8	2.65	2.16	42	$Phe_3Ser_5Lys_8$	>5.0	2.24

character in charge transfer complexes (1 for Trp, 0 for the other amino acids). σ_{4i} represents the electric charge of an amino-acidic side chain at pH= = 7 (EC). σ_{5i} represents the proton donor or acceptor character for hydrogen bonds (HB). As fractional atomic charges through conjugation or charges of ionic groups strengthen hydrogen bonds, σ_{5i} is considered ±0.75 for Ser and Thr, ±1.0 for AsN, GlN, Tyr, Trp and $\pm$ 1.25 for Asp, Glu, $HisH^+$, Lys and Arg. σ_{6i} pertains to dipole moments not involved in hydrogen bonds and is $+1$ if the significant dipoles of groups are approximately directed towards polypeptide chain (Trp) and —1 if they are directed towards the end of the side chain (Tyr; see section 3.3.6). In all cases σ_{ji} are the sums of absolute values of changes produced by all amino-acidic substitutions against the standard, for *j*th intermolecular force character. For each amino-acidic substitution in a given position of the peptidic chain, one takes the difference between the j-type increments of the amino acid in the compound i and of the one in the standard compound. The last column in Table 6.3 represents the minimum steric differences (MSD) introduced by the respective amino-acidic substitutions. Solving a system of linear equations as in Appendix 4, the best equation for the $N = 92$ compounds is:

$$A_i^{calc} = 0.45 + 0.12\sigma_{1i} + 0.68\sigma_{2i} + 1.1\sigma_{3i} + 0.004\sigma_{4i} + \\ + 0.35\sigma_{5i} + 0.77\sigma_{6i} \quad (6.7)$$

and the linear correlation coefficient $r = 0.75$, quite close to the —r values obtained with the simpler method in previous notes [13]*). As ΔG_{0i} (kcal mol^{-1}) $= 1.4 \log A_i$, the average decrease of the affinity for biological action (against the standard) caused by changes of σ parameters are (in kcal mol):

one unit difference in hydrophobic increment (σ_1) : 0.15
one unit difference in aromatic increment (σ_2) : 1.0
one unit difference in DCT increment (σ_3) : 1.5
a difference of one electrostatic (electronic) charge (σ_4) : 0.0
one unit difference in character of hydrogen bond (σ_5) : 0.5
one unit difference in dipole moment increment (σ_6) : 1.0

From these per unit decreases, proportional to the β_j values, the one related to charge-transfer complexes (β_3) is highly uncertain as σ_{3i} is non-zero for only two derivatives [15]. From the log A *versus* MSD correlation — a difference of one MSD unit causes a decrease of *ca.* 0.45 kcal mol^{-1} while for the enzymic hydrolysis data in Table 6.1 (section 6.2) — of *ca.* 0.60 kcal mol^{-1} [20].

Aromatic amino acids seem most important in determining the interaction specificity; the high β_2, β_3 and β_6 parameters corresponding to the AR, DCT and μ-characters correspond to them. Their importance may be also connected to a steric character — their planarity allowing a good contact with the partner-groups. One may observe that an appearance or disappearance of an aromatic cycle is accompanied also by a large MSD value. The low values of β_5 and especially β_4 may be explained by the polar and especially ionised groups remaining into contact with the aqueous surrounding even when the oligopeptide-biologic action site complex is formed. However there is an inverse relation be-

*) The mean standard deviation s=0.84, for a total A_i^{exp} range from 0.00 to 4.65.

tween the net electrostatic charges of the immunogenic macromolecules and those of the elicited antibodies [16]. This may be explained by the fact that our oligopeptides have a higher torsional freedom than proteins and other macromolecules. For proteins the polar groups of the combining sites may therefore be not always able to escape towards the aqueous surrounding when the antigen-antibody complex is formed.

Our differences in intermolecular force character generally present a certain parallelism to the minimal steric difference, and thus contain implicitely a steric character too. This explains why an attempt to introduce the MSD values as the seventh increment (as σ_{7i}) in equation (6.7) did not rise significantly the correlation coefficient r [20]. Thus, the real improving effect of introducing the type of intermolecular force over introducing steric differences is to rise r from 0.55 to 0.75. One may ask what determines the remaining 0.25 required to reach $r = 1$. The point is that we used increments, determined only by the amino-acidic substitutions, irrespective of position within the peptide in which the substitution occurs. The interactions between amino-acidic residues of the peptide are thus accounted for at most by a general mean value attributed to an amino acid, not to a certain interacting constellation of amino acids. A correlation of A_i^{exp} values for pairs of the same substitution with the mean A_i^{exp} values for the respective pairs gives a correlation coefficient $r = 0.79$ [22]. Thus, with increments for single amino acid substitutions in a linear equation of type (6.6), the best r value for peptide structure-activity correlations may be not higher than 0.75.

Table 6.4 illustrates an application of equation (6.7) to the correlation between the decrease of oxytocytic activity of 42 amino-acidic substitution derivatives (in rats; data listed by Sneath [12]) with intermolecular force parameters [20]. The chemical formula of oxytocyne is given in Figure 6.2; numbering of amino acid residues starts up from the H (amino) end. The $A_i^{exp} - A_i^{calc}$ correlation is characterised by $r = 0.74$ while for the A_i^{exp} — MSD correlation by $-r = 0.53$ [20]. As mentioned, if only substitutions in positions 3, 4 and 8 are considered one obtains $r = 0.95$ for the $A_i^{exp} - A_i^{calc}$ correlation and $-r = 0.70$ for the A_i^{exp}— — MSD correlation.

A structure-toxicity correlation attempt, based on parameters characterising intermolecular forces was done by Vîlceanu *et al.* [23] for phosphororganic compounds of the Schrader type:

$$\begin{matrix} R^1 \searrow & & \nearrow\!\!\!\nearrow O \\ & P & \\ R^2 \nearrow & & \searrow OR^3 \end{matrix}$$

The biological activity refers to toxicity for mice and rats (A_i^{exp} are log (molar) LD_{50} values). R^3 is the most acidic group which is eliminated when the compound phosphorylates the acetylcholinesterase — the probable mechanism of action for these compounds [24]. The figures for the parameters of the six types of intermolecular forces and for modified σ-Hammett constants were used (separately for $R^1 + R^2$ and R^3). The optimised linear equation, similar to equation (6.7), with eight adjustable coefficients, for 71 compounds, yields a correlation coefficient $r = 0.70$. As a general test for the validity of correlations with intermolecular force parameters, the absolute values of coefficients for the hydrophobic, aromatic and hydrogen bond parameters are similar to those of equation (6.7). The r value obtained in reference [23] is also low, but the equation encompasses a wide range of different chemical structures (the R^1, R^2, R^3 substituents), much wider than the structural range over which correlation equations, of the type used by Cammarata and Hansch, are usually valid (see section 4.4.3).

6.4. Specificity and molecular size

As mentioned at the beginning of this chapter, a certain relationship between specificity and molecular size should exist. A procedure of evaluating the size of the combining site (both on the recogniser and the effectors) will be outlined here.

The simplest reasoning is the one used by Thomas [2a] and by Conaughy and McCarthy [2c] in discussing the minimum size n of RNA oligonucleotides which can be recognised as self or alien, in hybridisation-competition experiments, by absorbed DNA (of viral, bacterial, *etc.* origin). These oligonucleotides are recognised by complementary base-pair formation with DNA segments. If the whole chromosomal DNA has the length of N nucleotides, there are also N overlapping oligonucleotidic sequences on this DNA; if their base sequences are randomly distributed, a too short RNA oligonucleotide coded by an alien DNA may have, by chance, a sequence complementary to those contained by the absorbed (self) DNA. If this is not to happen, the length n of the "alien" RNA oligonucleotide should be so that the total number 4^n of possible "words" in the base type alphabet with $\alpha = 4$ letters should be much higher than the number N of "words" on the "self" DNA:

$$\alpha^n \gg N; \quad n > \frac{\log N}{\log \alpha} \tag{6.8}$$

The point is that a single difference (B_i differing by a single "letter" replaced in the correct effector B_0) may not be sufficient to ascertain non-recognition by A_0, especially for high B_i and low B_0 concentrations. A very simplifield method to evaluate the size n of the combining site in the general case will be given in the next paragraphs.

6.4.1. *Notations and assumptions*

We shall start from equations (6.3) and (6.4), section 6.1. We need now some structural parameters to express the size of combining sites on the recognizer and effectors and the extent of structural modifications by which the combining site of B_0 differs from the one of B_i. The following notations are introduced:

n is the number of monomeric units in the combining site of the effector (B_0, B_i) species. The effectors are often polymers, oligonucleotides or oligopeptides, but one could also express the size n in term of atoms or second period atom-XH_n groups (in the case of MSD-parameter, section 6.2).

f_j; $j = 1, 2, \ldots, \alpha$ is the fractional content of the j-type monomer (atom) in the whole set of effectors species involved in the respective recognition process.

α is the number of monomer (atom) types within the effector species. The following relation is valid:

$$\sum_{j=1}^{\alpha} f_j = 1 \qquad (6.9)$$

ν_i is the number of substitutions (of monomers, atoms) by which the combining site of B_i differs from that of B_0

Δg is the mean affinity decrease (towards the recogniser A_0) produced by one substitution in the B_0 combining sites.

The assumption upon which our size calculation lies are [5a,e]:

(i) All effector recognition sites have the same size, n

(ii) The affinity decrease ΔG_i (of B_i towards A as compared to B_0) is proportional to the number ν_i of substitutions:

$$\Delta G_i = \nu_i \Delta g \qquad (6.10)$$

or, in relative association constants:

$$K_{0i}/K_{00} = r^{\nu_i} \text{ where } r \equiv e^{-\Delta g/RT} \qquad (6.10a)$$

(iii) The effector combining sites contain n well defined and not interchangeable positions in which the monomers (atoms) are placed.

6.4.2. *Calculation of size, n, of combining site*

Two different assumptions will be used for the distribution of the $[B_i]$ concentrations *versus* the number ν_i of substitutions:

(a) All B_i effectors differ from B_0 by the same number of substitutions within the combining site:

$$\nu_i = \bar{\nu}; \quad i = 1, 2, ..., N. \tag{6.11a}$$

(b) The concentration $[B(\nu)]$ of all B_i effectors differing from B_0 by ν substitutions is proportional to the probability $\pi(n, \nu)$ that a random array of length n, formed of α-monomer types, should differ by ν substitutions from the array corresponding to the B_0 combining site:

$$[B(\nu)] = [B_{tot}] \cdot \pi(n, \nu); \quad \nu = 1, 2, ..., n. \therefore$$

$$[B_{tot}] \equiv \sum_{i=1}^{N} [B_i]. \tag{6.11b}$$

Distribution (a) yields *via* inequality (6.3), for simple association-dissociation equilibria [5a]:

$$\bar{\nu} \geqslant \frac{2.3\,RT}{\Delta g} \log \frac{[B_{tot}]}{p_f[B_0]} = -\frac{1}{\log r} \cdot \log \frac{[B_{tot}]}{p_f[B_0]}, \tag{6.12}$$

while for recognition by precipitation governed by inequalities (6.4) [5d]:

$$\bar{\nu} \geqslant \frac{2.3\,RT}{\Delta g} \log \frac{[B_{mx}]}{[B_0]} = -\frac{1}{\log r} \cdot \log \frac{[B_{mx}]}{[B_0]}. \tag{6.13}$$

$[B_{mx}]$ is here the concentration of most frequent false partner B_i; log is the decimal logarithm ($\log_{10}$). This way one can calculate for distribution (a), the (mean) number $\bar{\nu}$ of substitutions by which the false effectors must differ from the correct one. Further on one requires that an accidental recognition by A_0 of a false B_i, belonging to the N random arrays, is small; in other words the probability that the combining site of any of the N false B_i effectors should not differ with more than $\bar{\nu}$ substitutions from that of B_0 should be small. If P is this probability one obtains (Appendix 5):

$$P \equiv 1 - [1 - p(n - \bar{\nu} + 1)]^N \cong Np(n - \bar{\nu} + 1) \ll 1 \tag{6.14}$$

$p(n - \bar{\nu} + 1)$ is the probability that an array of length n, of α-monomer types, should differ by at most $\bar{\nu} - 1$ substitutions from the combining site of B_0.

In case of distribution (b), if equation (6.10a) and (6.11b) are used in inequality (6.3) one obtains:

$$\frac{[B_{tot}]}{[B_0]} \sum_{\nu=1}^{n} \pi(n, \nu) \cdot r^{\nu} \leqslant p_f. \tag{6.15}$$

One assumes here $p_f \ll 1$. A relation based on somewhat similar principles was used by Sadler and Smith [17] for a theoretic size (length) evaluation of the *E. coli* lac operator. Distribution (b) cannot be applied directly to the precipitation recognitions as it refers to concentrations $[B(\nu)]$ of groups of false B_i-partners, while the solubility product relations (64) refer to each B_i partner separately.

6.4.3. *Quantitative evaluations and discussion*

For distribution (a) with inequality (6.14) and the result of Appendix 5 for $p(n - \bar{\nu} + 1)$, i.e.:

$$p(n - \bar{\nu} + 1) \cong \sum_{(jk)}^{(C_n^{n-\bar{\nu}+1})} \prod_{(k)}^{(n-\bar{\nu}+1)} f_{jk}, \tag{6.16}$$

one obtains an implicit equation for the calculation of n. If the relative contents f_{jk} of monomer types in all available B_i effectors do not differ much, one can approximate f_{jk} by $1/\alpha$ and inequality (6.14) becomes:

$$NC_n^{n-\bar{\nu}+1} \cdot \bar{\alpha}^{(n-\bar{\nu}+1} \ll 1. \tag{6.17}$$

By calculating the logarithm and using the Stirling approximation for X! the following implicit equation in n is obtained:

$$n > \bar{\nu} - 1 + \tag{6.18}$$

$$+ \frac{\log N + (\nu - 1) \log \dfrac{n}{\bar{\nu} - 1} - \dfrac{1}{2} \log \left[2\pi(\bar{\nu} - 1)\left(1 - \dfrac{\bar{\nu} - 1}{n}\right)\right]}{\log \left[\alpha\left(1 - \dfrac{\bar{\nu} - 1}{n}\right)\right]}$$

which may be compared with the more approximate result obtained in reference [5d]:

$$n > \bar{\nu} - 1 + \frac{\log N + (\nu - 1) \log \dfrac{ne}{\bar{\nu} - 1}}{\log \alpha} \tag{6.18a}$$

For distribution (b) the result for the $\pi(n, \nu)$ probability is (Appendix 5):

$$\pi(n, \nu) \cong \sum_{(jk)}^{(C_n^\nu)} (1 - f_{j1}), \ldots, (1 - f_{j\nu}) \cdot f_{j\nu+1}, \ldots, f_{jn}. \qquad (6.19)$$

With (6.19) and $f_{jk} = 1/\alpha$, inequality (6.15) becomes:

$$\frac{[\mathrm{B}_{tot}]}{[\mathrm{B}_0]} \sum_{\nu=1}^{n} C_n^\nu \left(\frac{1}{\alpha}\right)^{n-\nu} \cdot \left[r\left(1 - \frac{1}{\alpha}\right)\right]^\nu \ll p_f \qquad (6.20)$$

and if one calculates the logarithm (Appendix 6):

$$n \geqslant \frac{\log \dfrac{[\mathrm{B}_{tot}]}{p_f[\mathrm{B}_0]}}{\log \alpha - \log[1 + r(\alpha - 1)]}. \qquad (6.21)$$

If the composition of the B_0 combining site is known, i.e. the monomer types in the n loci of the site ($k = 1, 2, \ldots, n$) and also their relative contents f_{jk}, equations (6.18) and (6.21) may be utilised with an "effective" α evaluated as a geometric mean:

$$\bar{\alpha} = \left(\prod_{k=1}^{n} f_{jk}\right)^{-1/n}. \qquad (6.22)$$

Equations (6.18) and (6.21) for calculation of n contain a series of simplifying approximations which could not be endorsed but by comparison with experimental data. A first test: let all N partners have equal concentration and suppose a single monomeric substitution sufficient to prevent the recognition of false B_i from occurring ($\nu = 1$ that is Δg is large and r small). Both equations for n should in this case transform into Thomas' [2a] equation (6.8). For $\nu = 1$ equation (6.17) of which equation (6.18) derives, leads directly to equation (6.8). The same is true with equation (6.21) for $[\mathrm{B}_{tot}]/p_f[\mathrm{B}_0] = N$ and $r \ll 1/\alpha$. Likewise, n values calculated by equations (6.18) and (6.21) are not too different and depend the same way on N, α and r, as direct calculations have shown.

In the following chapter (sections 7.4.2, 7.5.2), the size of combining sites for the repressor-operator interaction *(E. coli* lac system) and for antigen-antibody interaction (antibody elicitation) are calculated and the results agree with the corresponding experimental data.

Qualitatively equations (6.18) or (6.18a) and (6.21) give the results intuitively predicted. The minimum size n of the combining site is the higher, the higher the number N of partners among which the recogniser A_0 must discern the correct B_0, and the lower the permitted error threshold, p_f. For the same N, p_f and Δg-values, the required size n is the lower the higher the number α of monomer types in B effectors.

6.5. Recognition with several selection steps

So far we have assumed that the selection of the correct B_0 out of the N types of effectors takes place at a single step. The requirement of sufficient specificity in such a process must be satisfied by a minimum size of the combining sites.

An alternative way, (cascade regulation), was suggested by Scherer [18], who assumed that selection may take place at several levels. Thus the process by which only certain proteins are synthesised by a cell (out of the total number encoded within the genome) may consist of a first selection step at the level of (approximate) repression of "unrequired" operons, a second step at the level of messenger RNA passing through the nuclear membrane in the cytoplasm, and a third at the level of an (approximate) selection by ribosomes of only the "required" messengers. A recognition process with two selection steps is shown in Figure 6.3. Step I consists of the formation of the A_0B_i complexes, with the probability p for the formation of the false A_0B_f complexes. Step II "controls" the complexes and dissociates the false ones, before they produce the false physiological action. The probability of escape of false A_0B_f complexes, or of a false dissociation for the correct A_0B_0 complex is p'. While step I is exoenergetic, step II may require some energy source, but this may not raise special difficulties. If φ is the total production rate of A_0B_i complexes, the rate at which the correct complexes reach the physiological action site

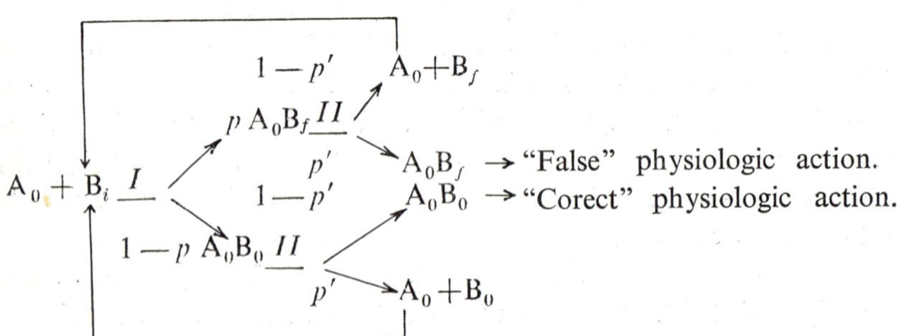

Figure 6.3. Cascade recognition.

is $\varphi(1-p)(1-p')$, while for the false complexes, the rate is $\varphi pp'$. The error probability P for the whole process is:

$$P = \frac{\varphi pp'}{\varphi(1-p)(1-p')} \cong pp'. \tag{6.23}$$

Thus the error P for a recognition process with s recognition steps will be equal to about the sth power of the error probability, p, of a single step:

$$P \cong p^s \tag{6.24}$$

Multiple selection will thus greatly lower the error probability if $p \ll 1$.

6.6. Recognition for several correct effector molecules

There may exist cases when recognition is correct for complexation of A_0 by M effectors, B_0, B_1, ..., B_{M-1} and false for the rest of $N-M$ effectors, B_M, B_{M+1}, ..., B_N. Intuitively, in this case would be required a degree of specificity lower than for the recognition of only one effector, B_0.

A lower specificity would really be required only if the combining sites of B_0, B_1,..., B_{M-1} are related in certain ways. Thus the combining sites, for A_0B_i-complex formation may be the same for all "correct" B_0, B_1, ..., B_{M-1} effectors. This case corresponds to a higher effective concentration, $[B_0]_{eff}$, of the correct effector:

$$[B_0]_{eff} = \sum_{i=0}^{M-1} [B_i] \tag{6.25}$$

while the number of false effectors would be somewhat lower, $N_{eff} = N-M$. The difference between N and N_{eff} may be insignificant.

Another case could be that the combining sites of the "correct" effectors differ with a relatively small number μ of monomer substitutions from the one of the "best" effector, B_0, such that $\mu < \nu$. The maximum number of recognisable effectors would be, in the sense of the assumptions made in section 6.4.1, N_{corr}:

$$N_{corr} = \sum_{l=0}^{\mu} C_n^l(\alpha - 1)^l \tag{6.26}$$

The maximum value of μ would be $\nu - 1$ with ν given by equations (6.12) or (6.13).

REFERENCES

1. E. Fischer, *Chem. Ber.*, **27,** 2985 (1894); P. Ehrlich, *Studies on Immunity*; J. Wiley, New York (1906); S. Arrhenius, *Immunochemistry*, MacMillan, New York (1907); quoted by A. Albert, *Selective Toxicity* (in Russian), Izd. In Lit., Moscow, (1953). John Wiley & Sons, New York, 1951.
2. (a) C. A. Thomas, jr., *Progr. Nucleic Acid Res. Molec. Biol.*, **5,** 315 (1966); (b) B. L. Conaughy and B. J. Mc. Carthy, *Biochim. Biophys. Acta*, **149,** 180 (1967).
3. M. V. Volkenshtein and A. M. Eliashevich, *Dokl. Ak. Nauk S.S.S.R.*, **136,** 1216 (1961)
4. D. F. Bradley, *J. Macromol. Sci. Chem.*, **4,** 739 (1970).
5. (a) Z. Simon, *J. Theoret. Biol.*, **9,** 414 (1965); (b) Cl. Nicolau, Z. Simon, *Molecular Biology* (in Romanian), Scientific Publishing House, Bucharest (1968). Ch. 5, §8; (c) Z. Simon, *Studia Biophysica (Berlin)*, **26,** 179 (1970); (d) Z. Simon, V. Gheţie, *Rev. Roumaine Biochim.*, **8,** 261 (1971); (e) Z. Simon, *ibid.* **9,** 337 (1972).
6. V. Gheţie, *Antibodies in action* (in Romanian), Encyclopedic Publishing House, Bucharest, (1971) Ch. 6.
7. (a) M. T. Palmiter and P. Aladjem, *J. Theoret. Biol.*, **5,** 211 (1963); (b) A. E. Nielsen, *Kinetics of Precipitation*, Pergamon Press, London, (1964).
8. A. D. Riggs, H. Suzuki and A. Bourgeois, *J. Molec. Biol.*, **48,** 67 (1970).
9. E. L. Kabat, *Structural Concepts in Immunology and Immunochemistry*, Holt, Rinehard & Winston, New York, (1968), Ch. 6.
10. M. Sela, *Naturwiss.*, **56,** 206 (1969).
11. D. Pressman and A. L. Groseberg, *The Structural Basis of Antibody Specificity*, Benjamin Inc. New York (1968), Ch. 3.
12. P. H. A. Sneath, *J. Theoret. Biol.*, **12,** 157 (1966).
13. (a) Z. Simon, Z. Szabadai and I. Miklos, *Rev. Roumaine Biochim.*, **8,** 79, (1971); (b) Z. Simon, *ibid.* **269** (1970); (c) Z. Simon, *Rev. Roumaine Biochim.*, **5,** 319 (1968)
14. (a) I. Schaechter, *Nature*, **228,** 639 (1970); (b) see reference [9] Table 6,8; (c) F. C. Wetall, A. B. Robinson, J. Caccam, J. Jackson and E. H. Eylar, *Nature*, **229,** 24 (1971); (d) K. H. Schleifer and R. M. Krause, *J. Biol. Chem.*, **246,** 986 (1971); (e) J. Spragg, R. C. Talamo, K. Suzuki, D. M. Appelbaum, K. F. Austin and E. Haber, *Biochemistry*, **7,** 4086 (1968); **8,** 3750 (1969); (f) E. Haber, F. F. Richards, J. Spragg, K. F. Austin, M. Valloton and L. D. Page, *Cold Spring Harbour Symp. Quant. Biol.*, **32,** 299 (1967).
15. Z. Szabaday and Z. Simon, *Rev. Roumaine Biochim.*, **9,** 327 (1972)
16. M. Sela, E. Mozes, G. M. Shearer, and Y. Karniely, *Proc. Nat. Acad. Sci. U.S.A.*, **67,** 1288 (1970).
17. J. R. Sadler and T. F. Smith, *J. Molec. Biol.*, **62,** 139 (1971).
18. Kl. Scherer, *Abh. Deutsch. Akad. Wiss. zu Berlin, Kl. Med.*, (1968) (1), 259.
19. S. T. Rao and M. G. Rossmann, *J. Molec. Biol.*, **76,** 241 (1973).
20. Z. Simon, and Z. Szabaday, *Studia Biophys. (Berlin)*, **39,** 123, (1973).
21. J. R. Knowles, *J. Theoret. Biol.*, **9,** 213 (1965).
22. Z. Simon, unpublished results.

23. (a) R. Vîlceanu, Z. Szabaday, A. Chiriac and Z. Simon, *Studia Biophys. (Berlin)*, **34**, 1 (1972); (b) *idem, Rev. Roumaine Biochim.*, **10**, 239 (1973); (c) *idem*, unpublished results.
24. R. D. O'Brien, in: *Chimie Organique du Phosphore*, Colloques Internationaux CNRS, Editions CNRS, Paris, (1970) p. 353.
25. R. W. Taft, Jr. in *Steric Effects in Organic Chemistry*, M. S. Newman, (editor), Wiley, New York, (1956), p. 568.
26. A. Dupaix, J. J. Bechet and C. Roucous, *Biochemistry*, **12**, 2559, 2566 (1973)
27. J. E. Amoore, G. Palmieri and E. Wanke, *Nature*, **216**, 1084 (1967); 27a. N. L. Allinger, *Pharmacol. Future Man, Pric. Int. Congr. Pharmacol.*, **5**, 576 (1972); R. A. Maxwell, (editor) Karger, Basel, 1973.

7. Analysis of some recognition processes

7.1. Enzymatic reactions

The enzymes catalyse reactions which involve micromolecules or monomeric units of polymers so that the active site of the enzyme interacts with a not too large number of substrate molecules. If an attempt is made to analyse, in the light of the previous chapters, the specificity requirements imposed on an enzyme in an organism on formation of enzyme substrate activated-complex (with the meaning of activated complex from absolute reaction-rate theory), the following conclusions are drawn. The overall number of micromolecular species from a living cell is not excessively large; so starting up with the building "bricks" one has: 20 amino acids, five bases, several kinds of monosaccharides, less than ten fatty acids. These, together with all intermediates produced on their synthesis and the existing micromolecular ions would not exceed a total of $N = 10^3$ micromolecular species, even if all groups on macromolecular surfaces in contact with enzymic sites are also counted. If counting is restricted to similar micromolecular species, like the number of nucleoside triphosphates among which a RNA- or DNA-polymerase must distinguish (for a certain nucleotide residue of the transcribed or replicated DNA strand), a $N \cong 10$ would come out. A maximum $p_f = 10^{-2}$ limit for the catalysis of reaction of one of the false partners, would probably suffice for precision. Let us try to apply equation (6.21) section 6.4.3 to the calculation of size n of combining site; this time, however, the size is expressed in atoms instead of monomer units. The number of types of the various atoms that should be considered is $\alpha \cong 10$. The average decrease Δg of the enzyme site-correct substrate affinity, for a substitution of atoms on substrate, should be of the order of a fraction of kcal mol^{-1}, thus $r \equiv \exp(-\Delta g/RT) \cong 0.5$. Consequently, with $[B_{tot}]/p_f\,[B_0] = N/p_f$ one obtains:

$$n > \frac{\log ([B_{tot}]/p_f [B_0]}{\log \alpha - \log 1 + r(\alpha - 1)} \cong \frac{\log 100N}{\log 10 - \log 5.5} =$$

$$= \begin{cases} = 12 \text{ for } N = 10 \\ = 20 \text{ for } N = 10^3 \end{cases} \qquad (7.1)$$

A "touching" of about 20 atoms from the substrate molecule by the enzymic site is probably sufficient in order that the enzymes "recognise" the correct reaction partners existing in the cell. However, a too high specificity in the interaction with synthetic micromolecules not existing in the cell, should not be expected, according to reasoning above. The number of molecules of a few tens of atoms which have been synthesised or are synthesisable in organic chemistry may easily exceed 10^6. Indeed, the iodine-acetate (ICH_3COO^-), considered as a specific inhibitor of glycerol-aldehyde dehydrogenase, sensibly decreases the activity of at least 55 other enzymes; the As(III)-compounds, considered by some authors as very specific inhibitors of keto-acid oxidation, at a concentration of 10^{-3} M, obstruct the activity of at least 52 enzymes [1]. Also, from among the four natural deoxyribonucleoside triphosphates (there are, in fact, eight if the ribonucleoside triphosphates are also considered) the DNA-polymerase chooses the correct partner with a probability to fail of only 10^{-8}—10^{-11} (the probability of spontaneous mutation per replication and base-pair [2]). Nevertheless, this enzyme can easily be "tricked" by addition of bromouracil, which incorporates in place of thymine and does readily pair, at the next replication, with guanosine triphosphate instead of adenosine triphosphate [3].

The notion of enzymic specificity gathers together four concepts [4]: (I) Reaction specificity, in the sense that only one type of reaction is catalysed; (II) Substrate specificity — only transformation of one or a few substrates is catalysed; (III) Binding specificity — binding of only one substrate to the active site of enzyme; (IV) Kinetic specificity refers to the difference in substrate reactivity of catalytic stages, subsequent to the initial binding to enzyme active site. The enzymes catalysing reactions with participation of biopolymers may have [5]: (I) primary specificity — the minimum structure required by the manifestation of enzymic specificity; (II) secondary specificity, which refers to the monomer units flanking the minimum structure, and eventually, (III) an overall spatial specificity, which depends also upon the rest of the biopolymer. For example, trypsin hydrolyses the peptide bonds whose acylic terminal is an Arg- or Lys-residue, while chymotrypsin does the same with aromatic amino acid—or Met-ended acyls (primary specificity); in case of chymotrypsine — as far as the residue next (towards the carboxyl-end) to the hydrolysed bond is concerned — Lys, Val and Ile favour, while Asp, AsN,

His, Ser and Pro obstruct hydrolysis (secondary specificity) [5a]. Studies involving the hydrolysis of the *p*-nitro-L-Phe-Phe bond in a series of oligopeptides reveal a marked influence of some more remote amino acid residues [5b] too.

The effect of spatial conformation upon enzymic reactions is enormous; let us only recall the fact that from among two optical antipodes or isomcrs (as substrates), generally only one is active in the enzymic reaction. This effect is, however, the most difficult to be expressed quantitatively. An interesting demonstration of the rôle of the steric factor in macromolecular catalysis is provided by the study of Lowrien and Linn [6], dealing with the protein catalysis of *cis-trans* isomerisation of some azo-dyes. Certain proteins increase by 1—2 orders of magnitude the rate of these isomerisations, while a variety of micromolecules, similar to protein side chains, or the same proteins, yet denatured by detergent — do not have any effect. The study of the spectra of these dyes (adsorbed on proteins) in solution shows that — in adsorbed state — these dyes are twisted about the —N=N— double bond (i.e they are brought into a configuration that is closer to the one corresponding to the activated complex for the *cis-trans* isomerisation).

The effect of intermolecular forces should appear especially in binding specificity. The maximum rate of several enzymic reactions is increased about 2—3 times by a single additional methylene in a hydrophobic and inert region of the substrate [7]. This increase corresponds to a variation of the free enthalpy of activation, $\Delta(\Delta G^{\neq})$, of *ca.* —0.6 kcal mol^{-1}, approximately equal to the hydrophobic increment of the CH_3-group (*cf.* Table 3.10, section 3.3.7).

From an analysis of kinetic data for hydrolytic enzymes, Knowles [4] concludes that a high hydrophobicity, together with the satisfying of electric charge conditions increases both the binding and the transformation specificity of the substrate. The active centre of chymotrypsine should be non-polar according to this analysis. Acetylcholinesterase, trombine and trypsine which split especially positively charged molecules, have their catalytic action usually increased towards substrates with large hydrophobic residues.

For other discussions concerning enzymatic reactions see sections 4.3.1, 4.3.2, 4.4.3, 6.2.1 and 6.3.

7.2. Accuracy ("Fidelity") in DNA replication and in protein synthetic processes

The accuracy of DNA-replication is given by the frequency of mutations per replication and per base-pair; it is of the order of 10^{-8} for bacteriophages, 10^{-10} for bacteria and 10^{-11} for primitive enkaryotes [2]. If discussion is restricted to nucleoside triphosphates, then the DNA polymerase must distinguish from eight types, the correct deoxyribonucleoside triphosphate monomers with afore-mentioned accuracy. In the synthesis of proteins there are several processes coupled in series and in parallel, which may be accompanied by errors. On one hand we have the transcription of DNA by RNA-polymerase, which should discern the correct (complementary to the transcribed position) ribonucleoside triphosphate among the eight types of nucleoside triphosphates. At transcription, the mRNA codons from the polyribosome complex should discern the correct aa-tRNA out of 64 aminoacyl-tRNA's (or, eventually, on average: 1 from 20). On the other hand, during aminoacylation of tRNA, each amino acid should complex with the correct transferase (from among at least 20), and the transferase-aminoacyl-AMP should aminoacylate the correct tRNA, from 64 that are present (or, on average: 1 out of 20, if the code degeneracy is taken into account). The error in the entire process of protein synthesis per amino acid residue was found to be smaller than 10^{-4} [9]. If the average polypeptide length is assumed to be *ca.* 10^3 amino acids, the above error limit is required in order that most of the synthesised polypeptide chains have the correct sequence. However, if some amino acid substitutions among similar (with respect to hydrophilicity/hydrophobicity ratio) amino acids on polypeptide chains do not affect the function of protein in near to 100% of the chain positions, then the level of error for amino acid incorporation could be lowered to 10^{-3}.

7.2.1. *DNA replication*

An accuracy of 10^{-8}—10^{-11} per base-pair per replication and the condition of discerning among even only four deoxyribonucleoside triphosphates, requires — in the sense of section 6.1 — a ΔG difference of 11—16 kcal mol^{-1} between the affinity of the DNA-base enzyme-site for the correct and false, deoxyribonucleoside triphosphate complexes, respectively. According to the discussion in section 4.2.1, for a false G:T pair, as the result of possible guanine tautomerism, one could not expect a ΔG difference larger than 5 kcal mol^{-1}. A hypothesis advanced by Lowdin (*cf.* section 4.2.2) has doubled this minimum value of ΔG. Another model of DNA-replications has been proposed by Kubitschek and Henderson [10]. This model assumes that the indirect precursors of DNA are pairs of deoxyribonucleoside triphosphates, paired by means of bases and in addition to the structural code of DNA, there exists a replication code for the selection of these bases by DNA-polymerase. The underlying basis for this replication code would be the 5- and 6-position groups of pyrimidine and positions 6- and 7-groups of purine. Namely:

CH_3;	O ... H_2N;	N	for the pair	TA
N;	NH_2 ... O;	CH_3	" " "	AT
H;	NH_2 ... O;	N	" " "	CG
N;	O ... H_2N;	H	" " "	GC

The replicase would be an allosteric oligomer with the configuration also depending both on the base-pair from the "old" strands which is to be replicated and on the replication code from the pair of monomers that is to be incorporated into the growing DNA-strands. According to this model, in order that a false base-pair A:C, C:A, G:T, T:G do present the same replication code as a correct pair, both bases must be in odd tautomeric and/or protonated forms. While, according to section 4.2.1., ΔG for the tautomerisation of cytosine and guanine may be smaller than 5 kcal mol^{-1}, for the tautomerisation of thymine and adenine, the ΔG-values are relatively high.

For example, the false pair of TG would have in the positions of replication code, the following groups: CH_3, O, O, N. By enolisation in positions 1—6 of guanine, this configuration becomes: CH_3, O ... HO, N and none of the "normal" could be reached, but through the protonation of OH in position 6 of guanine (although there would be a charge difference): CH_3, O ... H_2O^+, N — analogues to TA. For the false CA-pair, the configuration would be: H, NH_2 ... H_2N, N. By 1—6 isomerisation of the adenine, one obtains: H, NH_2 ... NH, N, which still has an additional proton over the CG-pair. For the protonation of the bases, the pK_a values are about 3 and 10. At pH = 7, the protonation and deprotonation ΔG is of the order of 5.5 kcal mol^{-1}, a value which adds to the tautomerisation ΔG and to which one may also add the decrease in pair affinity towards the enzyme, following the appearance of an additional positive and negative charge.

In vitro experiments indicate that DNA-polymerase extracted from leukaemic cells yield a higher level of erroneous incorporation than that extracted from normal cells. The levels of dGTP and dCTP incorporation into poly-(dAT), as catalysed by DNA-polymerase from human normal lymphoblasts, are 2.5×10^{-4} and 4.9×10^{-4} respectively, while for DNA-polymerase from leukaemic lymphoblasts, 5.2×10^{-3} and 2.1×10^{-3} respectively [79].

7.2.2. *The genetic code*

In the usual sense, the genetic code is the relation between the trinucleotide codons on mRNA (or on DNA-strand, complementary to the transcribed one) and amino acid coded by the trinucleotide. In Table 7.1 is given this genetic code, according to Caskey [9].

TABLE 7.1 **The genetic code**

Codons	Amino acid	Codons	Amino acid	Codons	Amino acid	Codons	Amino acid
UUU, UUC	Phe	UCU, UCC, UCA, UCG	Ser	UAU, UAC	Tyr	UGU, UGC	Cys
UUA, UUG, CUU, CUC, CUA, CUG	Leu	CCU, CCC, CCA, CCG	Pro	UAA, UAG	Term	UGA	Term
						UGG	Trp
				CAU, CAC	His	CGU, CGC, CGA, CGG	Arg
				CAA, CAG	Gln		
AUU, AUC, AUA	Ile	ACU, ACC, ACA, ACG	Thr	AAU, AAC	AsN	AGU, AGC	Ser
AUG	fMet, MET			AAA, AAG	Lys	AGA, AGG	Arg
GUU, GUC, GUA	Val	GCU, GCC, GCA, GCG	Ala	GAU, GAC	Asp	GGU, GGC, GGA, GGG	Gly
GUG	f Met, Met			GAA, GAG	Glu		

In fact, this code implies the following subcodes:

(a) in the transcription of DNA, based — at least for most part — on the formation of some complementary base-pairs;

(b) in the mRNA translation on polysome, based on the mRNA codon — aa-tRNA anticodon interaction, in most part also by means of complementary base-pairing;

(c) the amino acid synthetase (or amino acid-synthetase-tRNA complex) interaction, in the formation of amino acid adenylate (and eventually, also tRNA aminoacylation);

(d) the synthetase-tRNA (or tRNA-amino acid synthetase complex, during tRNA aminoacylation).

The first two subcodes imply oligonucleotide-oligonucleotide interactions, the third, protein-amino acid interactions, the fourth protein-nucleic acid interactions.

And now, before focusing on a discussion of the subcodes, a few general words about codes would be interesting [9]. The genetic code is universal, in the sense that the trinucleotide codon-amino-acid relationships are the same throughout the living world. From one class of organisms to another, the primary structures of tRNA and aa-tRNA (synthetases) differ, so that systems with tRNA from one class and with ribosomes and synthetases from another function with sensibly higher errors than systems with all components originating in one organism. There are, in a given organism, much more numerous species of tRNA (with respect to the codon-anticodon relationship) than the 20 amino acids; however the relative concentrations of various tRNAs are different in various tissues of the same organism. For example, in *E. coli* so far, 32 species of tRNA have been identified. The genetic code is degenerate, in the sense that most of the amino acids have several corresponding codons. In addition to amino acid codons, codons also exist for the initiation (fMet) and termination (Term) of polypeptide chain synthesis on polysomes; these are also universal features of the living world.

Genetic code degeneracy also seems to be necessary for the following reason: the 20 amino acids, plus termination and initiation (i.e. 22 signals) cannot be coded by dinucleotides, as with four bases one obtains only $4^2 = 16$ different dinucleotides; however with four bases one can form $4^3 = 64$ trinucleotides. One does not know much about how the present genetic code has been issued but it seems (from an analysis of its "logical" structure) that the most mutation-resistant code has been selected throughout evolution [9]. An analysis of monosubstitutions in codons (corresponding to point mutations) shows that in 25% of cases, the nucleotide substitutions do not lead to amino acid replacements, and in another 50% of cases the amino acid substitutions — although they actually occur — do not affect the polar — non-polar character of the coded amino acid [20]. It is interesting that the frequency of various amino acids, averaged over proteins of known primary structures, is approximately proportional to the number of codons corresponding to various amino acids [18].

As regards DNA-transcription, the complementary base-pairing (*cf.* Figure 5.4b, section 5.1) explains, at least qualitatively, the synthesis of RNA strand-complementary to the transcribed DNA-strand. However, the problem arises of signals by which the RNA-polymerase recognises the promoter region, the starting point of the operon as well as its terminal, i.e. the regions where DNA-transcription starts and ends, respectively. For some details see section 7.4. In Figure 7.1 the structure admitted for operon (according to [3] Figure 10, Chap. 10) is shown. **P** stands for promoter, whereby the RNA-polymerase transcription begins; OG the operator gene, where the repressor regulation takes place (*cf.* section 7.4); SG_1, SG_2, ..., SG_n are the synthesis genes, each one of which codes a protein. As regards transcription, the entire operon is transcribed, yet only the synthesis genes are actually translated (*cf.* [3], Chap.10); the DNA-strand transcription occurs in a 5→3 direction.

The translation of mRNA codons is conditioned by the first two bases; most of genetic code degeneracy refers to the third base; many amino acids are coded by four codons, with the third base being unimportant (Table 7.1). In order to explain this, Crick [21] advanced his "wobble" hypothesis, according to which only the recognition of the first two codon bases occurs by usual complementary base-pairing. The third position has some "freedom" — several bases might pair with a certain base of the tRNA anticodon. Thus, a G from the anticodon could pair to a U or C of the third position of mRNA codons; I (inosine) with U, C or A; C with G; U with A or G; and the modified uridine of tRNA with U, A or G. The tRNA anticodon is located on a loop at the oppo-

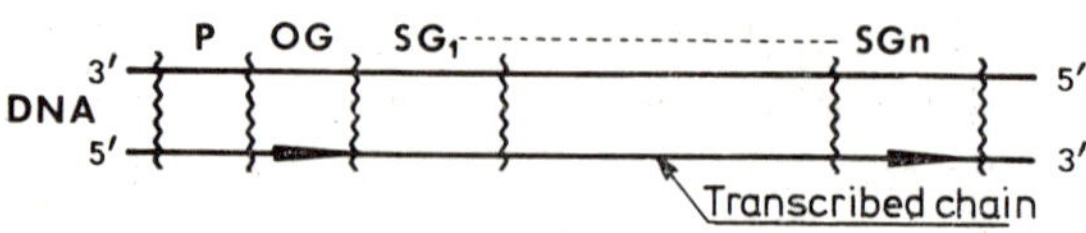

Figure **7.1**. Operon structure.

site end to the portion which binds amino acid (*cf.* Figure 5.7, section 5.3). The polysome codon-tRNA anticodon association constant is by far larger than that of the association codon trinucleotide-tRNA or codon trinucleotide-anticodon trinucleotide. *Explanation:* in the two latter associations there is a strong entropy effect, $\Delta S \ll 0$, of the fixing of structure in the complex, compared to the relative freedom of internal motion of the free trinucleotide. The codon in polysome and the anticodon in tRNA have a spatial structure already fixed, and thus the $\Delta S \ll 0$ effect disappears on association [45]. One special case is worthy of mention: the specificity of initiation and termination processes in protein synthesis cannot be accounted for only on the basis of complementary base pairing [9]. In elucidation of these processes, a great role has played the establishing of primary structure of phage R17 RNA [22]. This RNA has a length of 3300 nucleotides, but the very first translated "genes" appear only at about the 50th base from the 3′-terminal of the molecule. It is likely that initiation depends also on the secondary structure of RNA: in R17 RNA there is a double-helical region, just before the first AUG initiating codon; this codon is presumably exposed, due to its position, at the bending of a loop [22]. Otherwise, within the gene, the AUG and GUG codons are read as Met, which is incorporated into the growing polypeptide. For the initiation of protein synthesis, in addition to the fMet-tRNA Met complex and the suitable codon, certain protein factors are also required. Termination at *E. coli* takes place in the presence of UAA, UAG and UGA codons; in mammalians, the UAA was identified as the terminal codon [9]. The analysis of the primary structure of R17 RNA shows that next to the portion corresponding to envelope protein, there are two ending codons (UAA and UAG) and still an other 30 nucleotides before the AUG codon which initiaties the synthesis of the following (synthetase) protein. Two terminating codons — next to each other — could eventually increase the accuracy of termination. The recognition of termination codons is by a R protein factor, with two components = R^1 and R^2, which bind to ribosome in the presence of UAA or UAG, and UAA and UGA codons, respectively. Thus, R^1 and R^2 simulate the aa-tRNA molecules in their specificity for certain codons.

As regards the reactions of:

adenylation of amino acids (AA_i): $AA_i + ATP + E_i \rightarrow (AA_i - AMP \cdot E_i) +$ pyrophosphate;

tRNA aminoacylation: $tRNA_i + (AA_i - AMP \cdot E_i) \rightarrow AA_i - tRNA + E_i + AMP$, a model accounting for their specificity does not exist, as yet. Here E_i means the specific aa-tRNA synthetase for the aminoacylation of $tRNA_i$ with the amino acid AA_i. The interaction of $tRNA_i$ with an enzyme has been studied by cleavage of tRNA and by testing the affinity of various fragments towards the enzyme, or by chemical modification of tRNA at various positions and testing the same activity as above [23]. It is known e.g., that glycyl synthetase does aminoacylate several Gly-specific tRNAs [23b], as does seryl synthetase to several Ser-specific tRNAs [23c], which pleads for the non-participation of the anticodon region to enzyme recognition. In its turn, the deacylation of fMet-tRNA does not interfere with the binding of the latter to AUG codon (the dissociation constant remains the same 1.5×10^{-4} M) [23e]. Studies on tRNA fragments indicate the necessity of the anticodon double helix for synthetase recognition, yet the UH_2 and T loops (*cf.* Figure 5.7, section 5.3) do not seem to be directly involved in recognition. The splitting of yeast Val-specific tRNA at the internucleotide linkages of inosine (position 35) and guanine (position 57) yields inactive fragments if separated. On recombination, these fragments regain their affinity for valyl synthetase and may be aminoacylated by it [23a,d]. From among the theoretical hypotheses aiming to explain this specificity, some emphasise the stereochemical fit of the amino-acidic side chain to a portion of tRNA (however there also is a specificity in the interaction between amino acid and synthetase alone — *cf.* section 7.2.3.) [23g]. Other theories emphasise the specific interaction between synthetase and a specific nucleotide sequence of tRNA [23f] and suggest a trinucleotide code on RNA (differing from the anticodon region) for the recognition of various synthetases. This code does not involve unusual bases and is based on the hypothesis that the same relations between aminoaycl synthetases and tRNA's codons (for the recognition of these enzymes) with those between amino acids and their codons from the genetic code should be valid (see Table 7.2).

Some sequencing experiments of different tRNAs [80a,b] suggest that the recognition of the amino acid by the tRNA implies at least two steps. In the first step, somehow a rough selection would originate in a primitive recognition system. This first step implies the 4th base from the 3-end of the tRNA. Adenine selects for the non-polar amino acids, for Arg and Lys; guanine selects for Ser, Thr, Glu, Asp, GlN and AsN; cytosine selects for His and Pro, while uracyl codes for Gly [80b]. As to the initiation of protein synthesis in the initiator $tRNA^{Meth}$, the universal

TABLE 7.2 **Hypothetic synthetase recognition code**

First base	*Second base*				*Third base*
	A	C	G	U	
U	Phe	Ser	Tyr	Cys	G
A	Leu	Ser	—	Trp	G
U	Leu	Pro	His	Arg	A
A	Leu	Pro	GlN	Arg	A
U	Ile	Thr	AsN	Ser	C
A	Ile, Met	Thr	Lys	Arg	C
U	Val	Ala	Asp	Gly	U
A	Val	Ala	Glu	Gly	U

sequence —C—T—ψ—C is substituted by —G—A—U—C; possibly this tRNA does not function in amino acid incorporation [80c].

Considerations about how the codon-anticodon recognition mechanism operates were made by Ninio [81] on the basis of Crick's wobble hypothesis. According to Ninio's missing triplet hypothesis, this recognition is controlled by an enzyme which acts as a stereochemical filtre, verifying the geometry of the codon-anticodon complex. In order to avoid false recognition some of the possible 64 anticodons (on the tRNAs) should not exist and should therefore be missing.

7.2.3. *Errors in protein synthesis*

There is not too much information on the frequency of errors in the primary structure of natural proteins. Anyway, these errors are sufficiently small so that most proteins are functional and to sequence analysis a crystallised protein seems unitary. Nirenberg [11] assumes that most codons are translated with high accuracy — probability of error of 10^{-3}—10^{-4} or less, yet with certain codons this would reach 0.5 for certain amino acid substitutions. According to Ehrenstein [12], at least six positions out of the 141 of rabbit hemoglobin α-chain, would contain more than one amino acid, due to the variable translation of a unique template; in these multiplications only neutral amino acids seem to be involved. Data on false incorporations of nucleotides into RNA-strand at transcription do not exist (according to our knowledge). From the inhibition of poly-A synthesis (on some oligothymidylic sequences on *E. coli* DNA), the affinity of a site on transcribed DNA is 1000—2000 times larger for the complementary ribonucleoside triphosphate than for the non-complementary ones ($\Delta G = 4$—4.5 kcal mol^{-1}) [13]. There are some recent, more direct measurements of false amino acid incorporations in proteins. From the percentage of thermolabile glycosophosphate dehydrogenase in cultures of young fibroblasts a mean error level of 2×10^{-4} per amino-acidic position is inferred [82a]. Defectuous substitution of Val by Ile in rabbit reticulocyte hemoglobine has a frequency of 3×10^{-4} per residue; the codons of these residues differ only in the first nucleotide base [82b].

As regards protein synthesis in translating poly-U by the system extracted from rabbit reticulocytes, in addition to the 14-fold stimulation of Phe-incorporation, 5% Leu (against Phe) is incorporated (besides the missing poly-U) but incorporation of Lys and Arg has not been stimulated [14a]. In the system extracted from rat liver, poly-U stimulates erroneous incorporation of Leu (against Phe) at a probability of 0.25% at optimum Mg^{2+} concentrations [14b]. In experiments with labelled amino acids on hen oviduct protein synthesis, by isolating one by one the ovalbumins by enzymic hydrolysis and by verifying the uniqueness of the resultant peptides, it has been shown that Leu, Ile and Val do not substitute for each other (in ovalbumin synthesis) more than once in 3000 cases [14c]. This latter data, obtained from protein synthesis in the whole organ, would indicate a probability of erroneous incorporation of only about 10^{-4}; in fact, this study has dealt with only three amino acids, but the erroneous incorporation has been exceedingly low, given the similarity of these amino acids. Some antibiotics, such as streptomycin, lower the accuracy of translation and so do concentrations of Mg^{2+} ion, too high or low, or the utilisation of tRNAs and acylating enzymes from different organisms [9, 15]. The translation itself is not affected by combining the polysome system with aa-tRNAs from different organisms [9]. Streptomycin gets in at ribosomal level, causing mistaken reading of codon pyrimidines (U and C read as A) [15a]. The overall fidelity of translation of satellite tobacco necrosis virus ribonucleic acid in cell free *E. coli* system is rather low [83]; gel electrophoresis indicates a wide range of protein; only 5% thereof is real, correct coat protein.

There are studies on the accuracy of the tRNA aminoacylation process (with all components of the system originating in the same organism). The aminoacylation enzyme specific to Ile, in the absence of Ile-specific tRNA, distinguishes between Ile and Val to within an error of 4.5% in the reaction of coupling between the amino acid and an AMP residue; in the presence of Ile-specific tRNA, the error level drops at 0.5%. In the transfer reactions of an aminoacyl residue from AMP onto Ile-specific tRNA (catalysed by the same enzyme) the probability of transferring Val onto Ile tRNA is at most 0.02% [16a]. For the aminoacylation of tRNA (specific for Tyr, from *E. coli*) with L-Tyr: the Michaelis constant $K_M = 0.027$ mM, maximum aminoacylation rate of 2.6 μ mol mg^{-1} min^{-1}; with D-Tyr these values are of 0.015 mM and 0.11 Mmol mg^{-1} min^{-1} [16b]—a case of distinguishing isomer L from D with a high error ($p = 0.10$). In Table 7.3 are given the Michaelis constant (K_M) and the per

TABLE 7.3 **Precision of tRNA aminoacylation processes**

Process	*Enzyme*	*Substrate*	K_M(M)	k_2(%)	k_2/K_M(rel)
Phosphate exchange	$Val\text{-}tRNA^{Val}$ -synthetase (*E. coli* K12)	L-Val	6.7×10^{-5}	100	1
		L-Thr	7×10^{-3}	21	2.0×10^{-3}
		D,L-amino-butyric	1.5×10^{-3}	6	2.8×10^{-3}
	+ $tRNA^{Val}$	L-Ile	0.1	22	1.5×10^{-6}
Competitive substitution	$Phe\text{-}tRNA^{Phe}$ -synthetase (*E. coli* K12)	Phe	10^{-5}		
		Phe-butyl	8×10^{-5}		
		esther	6×10^{-6}		
		Benzyl alcohol	> 1		
		Gly	$>10^{-2}$		
		Tyr			
Aminoacylation	$Ile\text{-}tRNA^{Ile}$ -synthetase (*E. coli* B)	Ile + $tRNA^{Ile}$	7×10^{-9}		1.0
		Ile+$tRNA^{Phe}$	$>1.4\times10^{-4}$		8×10^{-4}

cent maximum rate (k_2) for valyl synthetase in the presence of Val-specific tRNA; they refer to AMP-ing of amino acid; the errors for tRNA amino-acylation might be smaller, as in the system studied by Loftfield and Eigner [16a]; other data [84a,b].

As regards aa-tRNA-codon interaction in polysomes, in a system entirely derived from *E. coli*, Arg-tRNA would recognise the UUU codon with a relative precision of 0.00 and so would do Ile-tRNA; Ser-tRNA recognises UGU codon corresponding to cystine (quoted by Caskey [9]) with $p = 0.03$.

In conclusion, since the overall error at codon translation—from DNA to protein—is very probably betwen $10^{-3} — 10^{-4}$ for most of codons, for none of the processes involved in translation, does the error exceeds this level. In DNA-transcription, in order that RNA-polymerase be able to discern the correct ribonucleoside triphosphate out of four false ones, the difference between the affinities of correct and false nucleotides (for the site of transcribed DNA + enzyme), should be of the order of $\Delta G =$ 5—6.5 kcal mol^{-1}. The method fails with tRNA aminoacylation: each amino acid has to distinguish among at least 20 tRNA-enzyme complexes with an accuracy of $10^{-3}—10^{-4}$. Admitting the interaction of the amino acid with both enzyme and tRNA bases (odd bases included) we come out with 30 types of monomers in tRNA-enzyme complexes. Applying the same reasons as in the previous chapter to calculations of the number n of monomers in the combining site of complexes with which the amino acid has to interact, we get (with $N = 20$, $p = 10^{-3}...10^{-4}$, $\alpha = 30$, $\Delta g =$ 2.0 kcal mol^{-1}; *cf.* section 6.3) an average value $\nu = 3—4$ (equation (6.12), section 6.4.2). With equation (6.18a) (section 6.4.3) one obtains $n = 6—7$, and with equation (6.21) (section 6.4.3) $n = 4—5$. Thus, each amino acid should interact with approximately five amino acids or bases from the tRNA-aminoacylating enzyme complex, which is not impossible, especially with the larger amino acids. Anyway, one should expect relatively high errors in the translation of codons for Ala and especially Gly at the level of aminoacylation of the corresponding tRNAs. However, one should bear in mind that the deduction of formulae for the size n of the combining site in Chap. 6, was based on some statistical considerations, which do not apply to tRNA aminoacylation, where we have to do with small values for N and n. Thus, a small locus and a high rigidity of Gly-specific aminoacyl transferase could ensure a very low error in translating codons for Gly. Nevertheless, aminoacylation of $tRNA^{Ala}$ by Gly, instead of Ala, implies as $\Delta G^{\neq}$ losses only the hydrophobic increment of the CH_3 group (missing in Gly) and that of an empty hydrophobic cavity which accommodates (in correct aminoacylation) the CH_3 group of Ala. The affinity difference, ΔG_i, in this case could not be more than *ca.* 1.5 kcal mol^{-1} and the error probability in aminoacylation of $tRNA^{Ala}$ by Gly, in a single-step recognition, would be $p_f \sim 0.1$. The same would be true for aminoacylation of $tRNA^{Ile}$ by Val.

From the viewpoint of the size of recognition site, the DNA-transcription and mRNA-codon—tRNA anticodon interaction raise fewer problems than the correct aminoacylation of tRNA. If the Watson-Crick mechanism of base-pairing could hardly ensure the $\Delta G = 5$—6.5 kcal mol^{-1} difference in affinity during transcription (owing to, especially, the readily enolisation of guanine and cytosine (*cf.* section 4.2.1), a mechanism of enzymic "touching-feeling" of the base-pair, of the type advanced by Kubitschek and Henderson (section 7.2.1) easily solves the specificity requirements. In the codon-anticodon interaction, it would be tempting to identify the affinity decrease, ΔG_i, with the difference ΔU_i, of interaction codon-false anticodon and codon-complementary anticodon, according to Claverie's calculations ([17] and also section 3.1.1.). However, this decrease of ΔU_i does not supply enough accuracy: $\Delta U = U$(UGG: CCC) - U(UGG: ACC), corresponding to the reading of the only Trp-codon as Gly, is only 3.0 kcal mol^{-1}, i.e. the probability of error is $p = 4\times10^{-3}$, and ΔU for the reading of the two Cys-codons as Gly is only *ca.* 2.1 kcal mol^{-1}, corresponding to a $p = 3\times10^{-2}$ [18]. According to Woese [19], the ribosome would leave a trace in the accuracy of translation. An analysis of ΔU calculated values in correlation with the corresponding amino-acidic substitution [18], shows that for many cases with small ΔU-values corresponds an important change in the types of intermolecular forces, or, even a transition from a non-polar character to a polar character of the amino acid. For example, ΔU for the interaction of CUG with GCC (against GAC), corresponding to the subsitutions of Arg for Leu, is only 3.3 kcal mol^{-1}.

In conclusion, tRNA aminoacylation, as indicated both by figures of Table 7.3 and by the theoretic considerations for substitution of Ala by Gly and of Ile by Val, could hardly meet the required error probability of $p_f \cong 3\times10^{-4}$ for false amino acid incorporation in the whole protein synthesis. There may be some supplementary controlling recognition processes after tRNA aminoacylation, for example at the level of ribosoma translation, as in the "cascade" regulation theory of Scherer (*cf.* section 6.5). Also, some substitutions, as Ala by Gly or Ile by Val may not change shape and function of proteins. Rapid degradation of at least some falsely-shaped proteins may intervene also in increasing the precision of the synthesised functional protein: an *Escherichia coli* mutant which produces unfinished (shorter) chains of lac repressor or β-galactosidase degrades rapidly these unifinished chains. The relatively rapid degraded 2% of the synthesised protein in the wild type may represent such a "corrective degradation" [85].

Errors in protein synthesis play a crucial rôle in Orgel's theory of aging by error accumulation [86] (producing increasing quantities of defective DNA polymerase and thus increased mutation frequency in the following cell generation—the "error catastrophy"). Prolonged hyperfunction produces degeneration of the respective organ—see for example Arcos *et al.* [87]. It would be of interest to see how hyperfunction conditions could enhance errors in protein synthesis. The cell aparatus in hyperfunction will have few abilities to produce highly endoenergetic molecules or to absorb, by active transport, species not required in the hyperfunction process. The intracellular concentrations of amino acids which are relatively rare in blood may thus decrease and the probability of false tRNA aminoacylation (for example) by more common amino acids will increase, *cf.* equation (6.3), section 6.1.

7.3. Competition-hybridisation among nucleic acids

The competition-hybridisation experiments with nucleic acids have been carried out as follows. On a solid carrier is adsorbed a macromolecular nucleic acid, e.g. denatured DNA extracted from a given organism. This adsorbed DNA is placed in contact with a solution containing oligonucleotides of a certain average size, originating in the cleavage of some RNAs from various sources, or with these same non-adsorbed RNAs *per se*, on the solid carrier. From among these RNAs one is radioactively labelled, and, therefore, its adsorption on DNA (and, consequently, on the solid support too) can be quantitatively determined. With such kind of systems it is supposed that the RNA or the oligonucleotide species complementary to a certain DNA-sequence do replace all the others on this sequence. Thus, one may specifically evidence a certain type of RNA from a mixture; e.g. the *E. coli* RNA will replace any RNA (from other sources) from the *E. coli* DNA. Also, setting into competition RNAs from various tissues and RNA synthetised *in vitro* by deproteinised (i.e. entirely transcribed) DNA, one can evaluate for the DNA extracted from the organism dealt with, the transcribed percentage of this DNA in the respective tissue [25, 26].

This type of experiment is relevant hereby in the following respects. What is the minimum length the oligonucleotides should have in order that they be distinguished (from among a mixture of oligonucleotides) by that DNA sequence which synthetised them (that is a complementary sequence)? Does a one base difference in oligonucleotide sequence suffice for it to be recognised as non-complementary? Long oligonucleotide sequences may assume a multiplicity of spatial configurations and their covering on interaction with DNA may require a very long time interval. Therefore in conditions of competition-hybridisation experiments, is any equilibrium reached, or if not, what influences may the departure from equilibrium have?

The first question has been answered by hybridisation experiments with DNA extracted from various organisms and with oligonucleotides differing in size, originating in the same organisms as the DNA, or in different ones. Thus, phage T4 DNA can distinguish the oligonucleotides coming from RNA yielded by even T phage, from those coded for by T odd phage, provided that the oligonucleotide length is at least of 10 monomers [27]. In order that DNA of *E. coli* or of mammals be able to distinguish the self-oligonucleotides from the non-self ones, the minimum length should be of 12—17 monomers [26].

If a single base change is sufficient for an oligonucleotide to be recognised as non-complementary, this corresponds to the particular case of $\nu = 1$ from section 6.4. In this case, the reasoning offered by Thomas [25] and by Conaughy and Mc Carthy [26a] is as follows. In order that an n-length oligonucleotide recognises its gene corresponding to the RNA wherefrom the oligonucleotide comes, on a single-stranded DNS of length N, only one of the $N—n \cong N$ successive sequences of n bases on DNA should be complementary to the oligonucleotide. Since four bases may yield 4^n different oligonucleotides, one must have $4^n \gg N$ (*cf.* equation (6.23), section 6.4.3):

$$n > \log N/\log 4 \tag{7.2}$$

The phage T4 DNA has $N = 4 \times 10^5$ nucleotides; thus, equation (7.2) requires $n > 9$; the *E. coli* or *Salmonella typhimurium* DNAs have $N = 4 \times 10^6$, requiring $n > 11$; the DNA of the haploid complement of mammalian genome DNA has $N = 5 \times 10^9$, requiring $n > 16$. These values are in agreement with the experimental lengths of oligonucleotides; over these lengths appears the species specificity [26a].

These competition-hybridisation techniques for denatured DNA attached to a filter, hold in principle, for unfractionated RNA (which is anyway far smaller than DNA) too. However studies of this kind show that with increasing oligonucleotide lengths, the time required for reaching an equilibrium state strongly increases. Consequently, for oligonucleotide sequences longer than $n \cong 20$, the equilibrium is practically never attained, non-complementary sequences [28] remaining fixed on DNA. The study of kinetics of this competition-hybridisation process reveals however that whenever excess RNA is employed, there exists a linear relationship between the reciprocal of the hybridised RNA amount per μg DNA and the reciprocal of time of contact t, no matter what the amount of DNA and the degree of cleavage of RNA. If $t_{1/2}$ is the time required by reaching of 50% of saturation hybridisation, the $C_r \cdot t_{1/2}$ product increases approximately linearly with the complexity of the RNA base-sequence which is hybridised. Here C_r is the molar concentration of RNA nucleotides in solution. That is, the hybridisation rate depends on the concentration of oligonucleotide sequences, each complementary to a DNA base-sequence. The more species of RNA molecules in solution, the lower the concentration of each oligonucleotide sequence. Under certain conditions, by comparing the hybridisation rate to a standard kinetic curve, one can calculate the number of RNA species in the solution dealt with [29b].

7.4. Repressor-operator interactions

In the regulation of DNA-transcription, an important part is assigned to the interaction between the repressor and the operator gene, inhibiting the operon transcription by the RNA-polymerase attached to the promoter gene (see Figure 7.1), and the interaction between RNA-polymerase, promoter gene and promoter, which favour the binding of RNA-polymerase to promoter gene and operon transcription (negative and positive control, respectively; [3], Chap. 10). These control systems have been extensively studied with *E. coli*. Till now, the following items have been isolated: the *lac* operon repressor (which codes the enzymes for lactose actively entering the cell and its splitting into glucose and galactose) by Gilbert and Müller-Hill—the so-called *lac* repressor [30]; phage λ repressor, by Ptashne [31]; an arginine repressor [32]; the galactose operon repressor (*gal* repressor) [33]; and promoters *ara* C protein, necessary for the expression of the L-arabinose operon [34], and the CGA protein, required for the activation of the *lac* operon [35]. It seems that the so-called σ-factor of RNA-polymerase protein has a specific contribution in transcription [36]. These repressors and inductors are oligomeric proteins, free of RNA, in quantities of the order of 10—40 molecules cell^{-1} (for a volume $v = 10^{-12}$ cm^3

in a concentration of the order of $10^{-7} \ldots 10^{-8}$ M) [37], and they specifically interact with certain portions of DNA; the dissociation constant is *ca.* 10^{-12}—10^{-13} M. Thus the problem arises of specific interaction between DNA and proteins, which is discussed here to illustrate the interaction between the *lac* repressor and the corresponding operator gene; this is by far the most studied example among the known repressors and promoters.

From other specific protein-nucleic acid interactions thus far known, we mention that the termination factors—the R_1 and R_2 proteins—recognise UAA and UAG, and UAA and UGA respectively, on the polysomal mRNA. Ribonuclease A recognises U or C from RNA while ribonuclease T_1, G or C; the affinity of poly-Arg towards nucleotides decreases in the order: GMP $>$ AMP $>$ CMP $\gg$ UMP (data quoted by Caskey [9]).

We mention that the affinity of repressor for the corresponding operator genes depends strongly on the pH and ionic strength of solution in which the respective measurements are made, as well as on temperature and the various *E. coli* strains from where the repressor or DNA was extracted. Mutations in the regulator gene or in the operator gene strongly influences this interaction process as well as the interaction of repressor (or of repressor-operator complex) with micromolecules which have a weakening (inductors) or strengthening (co-repressors) rôle in repression.

7.4.1. *Data concerning interaction between* lac *repressor and* lac *operator.*

The *lac* repressor is a tetrameric protein, with a global molecular weight of 150 000 dalton (4 units of 40 000 dalton each). The monomer has an overall composition of: Lys_{13} His_6 Arg_{17} Asp_{32} Glu_{44} Thr_{19} Ser_{26} Pro_{14} Gly_{16} Ala_{41} Cys_3 Val_{29} Met_8 Ile_{20} Leu_{38} Tyr_8 Phe_8 Trp_2. Thus the negatively charged amino acids prevail at pH $= 7$ [38]. Fifty amino acids in this repressor (sequenced by Müller-Hill *et al.* [69]) from the NH_2-terminal seem to be involved with the binding to operator gene. The *lac* operator has a certain symmetry, being formed of two equivalent segments with five or eight sites, and arranged as below [39a]:

VIa—Va—IIa—IIIb—IVb—IVa—IIIa—IIb—Vb—VIb.

The following hypothetic sequence for the operator gene has been proposed [69]:

$$\left.\begin{matrix} 5' \\ 3' \end{matrix}\begin{cases} \text{C,A—T—A—G,T—C,A—(X)}_n\text{-} \\ \text{G,T—A—T—G,A—G,T—(X}^{-1})_n\text{-} \end{cases}\right]_4 ;\ \text{probably, } n = 1$$

X stands for an unknown base, and C,A—for example—means that in the respective positions one finds either a cytosine or an adenine.

From among these sites, Ia and Ib are apparently formed (according to [39a]) from two AT pairs, sites IIa, IIIa, IVb, Vb and VIb of an AT-pair, IIb of three pairs of GC, and IIIb and IVa of GC-pair. The operator size, based on some statistical calculations of frequency analysis of recombinations, is of 10—50 base-pairs [39b], and the length of *E. coli* DNA segments, protected against DNAse action by binding to *lac* repressor, is of 20—23 base-pairs (Gilbert, quoted by [38]).

The *lac* repressor-operator interaction has been thoroughly studied from both thermodynamic and kinetic points of view by Riggs *et al.* [40]. Employing a highly sensitive membrane-filter technique and tracer compounds, these authors show that the association-dissociation of repressor obeys a simple rule:

$$\frac{\mathrm{d[RO]}}{\mathrm{d}t} = k_a([\mathrm{O}] - [\mathrm{RO}])([\mathrm{R}] - [\mathrm{RO}]) - k_d[\mathrm{RO}], \tag{7.3}$$

where k_a and k_d are the association and dissociation rate constants, respectively; [O], [R], [RO] are the concentration of the operator gene, free repressor and the repressor-operator complex concentrations, respectively. The association rate constant is of 7×10^9 M^{-1} s^{-1} and decreases with the increase of ionic strength (corresponding to a product of charges of $Z_1Z_2 = -6$); the association rate constant calculated from diffusion data is of 10^8 M^{-1} s^{-1}. The increase in association rate should be due to electrostatic attraction, and not to an association of repressor somewhere on DNA, followed by its sliding towards the operator. This is so because the cleavage of DNA does not influence the association kinetics while the dissociation equilibrium constant of a repressor from the remaining part of *E. coli* DNA is at least 10^{-3} M; the repressor does not associate with non-operator DNA at intracellular concentrations. For the dissociation of repressor off the operator, the equilibrium constant in intracellular conditions is $K = k_d/k_a = 1 \times 10^{-13}$ M, that is the dissociation rate constant, $k_d \cong$ $\cong 0.05$ min^{-1}, increases with ionic strength. For the association-dissociation equilibrium at 24 °C, $\Delta G = -18$ kcal mol^{-1}, $\Delta H = +8.5$ kcal mol^{-1}, $\Delta S = +90$ kcal mol^{-1} deg^{-1}. Therefore the motive force is the increase in entropy, possibly by means of the release of hydration water.

The specificity of repressor-operator association is high. The repressor does not bind to denatured operator DNA, yet it does after renaturation. A 300-fold excess of *E. coli* DNA without operator gene does not compete sensitively with DNA containing the operator gene [39a]. Since we have $N = 4.5 \times 10^6$ short oligonucleotide sequences on *E. coli* DNA, $[B_{tot}]/[B_0] \cong 1.5 \times 10^9$ (equation (6.11b), section 6.4.2); 4.5×10^6 is the DNA length in nucleotide-pairs. By combining equations (7.3) and (7.5) (with $p_f = 0.1$; section 6.1), the average value for the ΔG decrease in affinity of repressor for *E. coli* DNA — against *lac* operator—is $\Delta G \cong 14$ kcal mol^{-1}, or $G_{false} \cong 4$ kcal mol^{-1}, a value which would correspond to the above mentioned $K \cong 10^{-3}$ M.

The mechanism proposed by Gierer [41] according to which the specificity of DNA-protein interaction is due to a particular spatial structure (like the tRNA three-dimensional structure) of a DNA-portion—is unlikely [39c], because the binding of repressor to renatured DNA indicates that the operator is in a thermodynamically stable state (very likely a double helix). Also, the association rate which is even higher than the diffusion rate precludes a conformational transition during repressor-operator interaction.

The *lac* repressor has a high affinity for poly-dAT, $K = 10^{-9}$ M; its affinity for polynucleotides decreases according to the sequence: poly-dAT > poly-dA:dT > poly-dT > poly-dG:dC. With these polynucleotides, too, the affinity decreases with increasing ionic strength, demonstrating the participation of some electrostatic attractions (phosphates-basic amino acids) in the repressor-operator interaction [42].

Point mutations in the *lac* operator lower its affinity for repressor by 8—100 times [39a]. It is unusual that only one base-pair substitution (out of three possible) occurs or all the three with the same probability [39b]. The geometric mean of affinity decrease over the six classes of point mutations discovered so far [39a], yields an average r (*cf.* definition section 6.4.2) of 0.027, or the mean decrease of affinity per base-pair substitution, $\Delta g = 2.1$ kcal mol^{-1}.

The wild type lac repressor is not the one with the highest known affinity for the lac operator gene. The constitutive lac *E. coli* mutant [86] produces a repressor whose dissociation constant for the repressor-operator interaction is 10^{-15} M; i. e. this mutant-repressor has an affinity for the operator about four times higher than the wild type repressor [88].

7.4.2. *The size of combining site*

The size of the combining site, i.e. of the *lac* operator, has been calculated by mean of formulae (18a), (section 6.4.3), yielding $n \cong 22$ to 35 base-pairs [43]. Another calculation performed by Sadler and Smith [39b] yielded $n = 18$, by assuming that all the non-operator sequences differ from the operator sequences by at least three base-pairs, and by employing statistical reasoning similar to that in Appendix 4, which leads to evaluation of probabilities $\pi(n, \nu)$.

Starting with $N = 4.5 \times 10^6$ partners, $[B_{tot}]/p_f[B_0] = 1.5 \times 10^{10}$, $r = 0.027$ or $\bar{\nu} = \Delta G/\Delta g = 7$ and $\alpha = 4$ types of nucleotide pairs, by applying equation (6.18) or (6.21), section 6.4.3, one obtains $n > 26$ or $n > 18$ respectively.

According to Gilbert and Maxam [98] the *lac* repressor protects from deoxyribonuclease digestion a fragment of DNA double helix of about 27 base-pairs long. This should be the length of the lac operator, in good agreement with the above mentioned calculated value [43].

Maniatis and Ptashne [89] determined the size of the λ-bacteriophage operator. This operator consists of six identical repeated combining sites for repressor molecules. The size of the operator fragment which binds the first repressor dimer is *ca.* 35 nucleotide pairs. Specificity requirements here should be about the same as for the *Escherichia coli* lac system: the λ-bacteriophage gene is included in the bacterial genome which codes also for the repressor.

7.4.3. *Intermolecular forces and complementarity relations in the DNA-repressor interaction*

The first proteins considered as repressors were the histones [45] which make up a high percentage of chromosomal proteins and which are basic. Thereafter complexes built of chromosomal proteins coupled covalently with short RNA molecules (50—100 nucleotides), found also in chromatine were considered for this rôle [46]. If the histones present an extremely reduced specificity in inhibition of transcription [46], the chromosomal RNA, to whom the specificity of inhibition was attributed is very likely to be an artefact [47]. Thus, in eucaryotes, the regulation of DNA transcription is not cleared up, but it should present at least as much specificity as the one in procariotes and it implies most probably also specific DNA-protein interactions.

A recent review of the mechanisms which may assure the specificity of protein-DNA interactions was given by Hippel and McGhee [75]. Native DNA may well exist in solution in a family of conformations resembling the B-form, but differing in topological detail depending on local base sequence and composition. Generally d(A-T) rich DNA is more easily deformed from the B-form (see section 5.4). Such local topology differences may alter somewhat the distances between the negatively charged phosphate residues and thus offer some spatial differences for interactions with multiple positive residues of proteins [75].

There are hydrogen bonding determinants at the edges of the stacked base-pairs which distinguish dA.T from dG.C. base-pairs. The situation is illustrated in Figure 5.4b, Chap. 5, in which the wide groove is upwards from the shallow groove, downwards towards the base-pairs. In the wide groove, contact may be made with substituents located on the 6-, 7- and 8-positions of the purine and the 4-, 5- and 6-positions of the pyrimidine, while the shallow groove provides access to position 2 on the pyrimidine and 2 and 3 on the purine. Thus, in the wide groove, the dTA pair presents the functional group sequence $(CH_3)^5$, $O^4 \ldots HN^6$, N^7, the dCG pair the sequence H^5, $(NH)^4 \ldots O^6$, N^7. Figure 7.2 presents these functional groups on the DNA base-pairs as well as some pertinent distances. Figure 7.2b presents hydrogen bonding groups on amino-acidic side chains, as well as spatial conformations and distances which allow formation of two hydrogen bonds with the amino-acidic side chains: as argued in section 3.3.5, the second hydrogen bond may give a sensibly higher contribution to the binding free enthalpy than the first. On this basis some amino acid *versus* dAT or dCG base-pair complementarity relations may be searched [50]. Double hydrogen bond formation seems possible between (a) the amidic group (AsN, GlN) and $(NH_2)^2$, $(N)^3$ of guanine in the dCG pair; (b) the ammonium cation (Lys) and $(O)^2$ of thymine and $(N)^3$ of adenine in the dAT pair, but with some distortion; (c) the guaninic cation (Arg) with $(O)^6$ and $(N)^7$ of guanine in the dCG pair; (d) the amidic group (AsN, GlN), with $(NH_2)^6$ and $(N)^7$ of adenine in the AT pair. The preferential binding of poly-L-lysine to dAT rich DNA [48] may be explained by the double hydrogen bonding of type (b) [8] as well as by the more easy deformability of dAT-rich DNA and the hydrophobic interaction between the methyl group of thymine and the tetramethylenic moiety

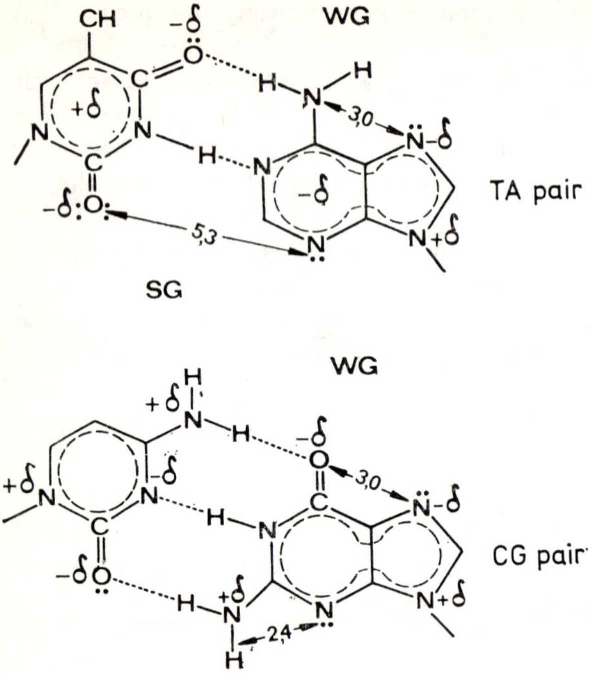

Figure 7.2. Interatomic distances and charge distributions in the base-pairs AT and CG.

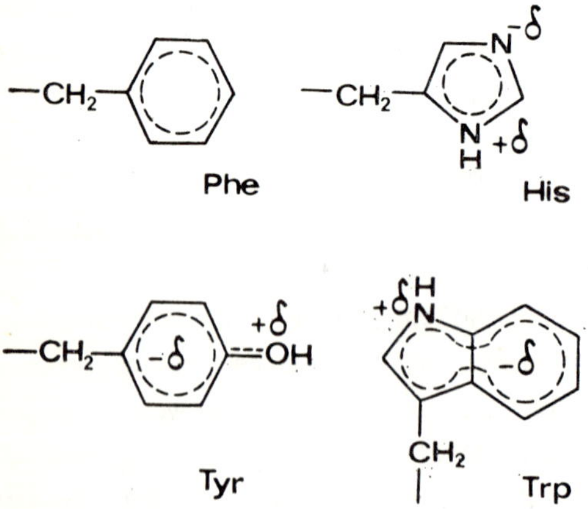

Figure 7.2. (a) Hydrogen bonding groups, interionic distances and approximate charge distribution in dAT and dGC base-pairs (distances in Å) (*cf.* section 3.3.6 and Figure 2.4., Chap. 2)

WG = wide groove. SG 1 = shallow groove

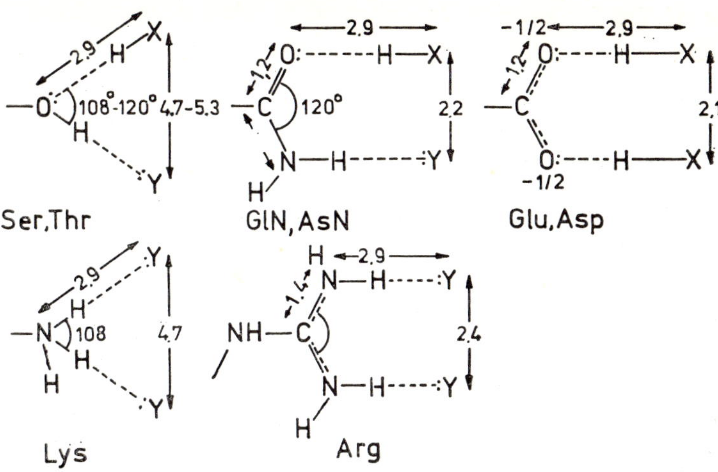

Figure **7.2** (b) Distances (Å) and orientations required for double hydrogen bonds between amino-acidic side chains and proton donating (X—H) and accepting (—Y:) groups.

of lysine [51]; nevertheless poly-L-Lys also shows preference to the dAU double helix [75]. The intervention of hydrogen bonds in the specific poly-dAT-poly-L-Lys binding may explain why this specificity is only obtained at sufficiently high ionic strength: ammonium Lys groups become available only if some ionic bonds of ω-ammonium (Lys)-phosphate residue are dissociated. Poly-L-Arg also shows some weak specificity for dCG rich DNA [49] which may also be explained by double hydrogen bond formation of type (c) with arginine side chains forced into the DNA wide groove.

Another mechanism for specific DNA-protein interactions may be based upon specificity of the aromatic amino acid side chain intercalation in the DNA double helix [50] (*cf.* section 3.3.8). Aromatic hydrocarbons and dyes such as acridine or ethidium bromide may intercalate quite easily in the DNA double helix with a rule favouring the overlap between the hydrophobic side planes of aromatic cycles [44]. Thus for intercalation of acridine in DNA, unwinding of only 12° of the double helix is required, which enlarges the distance between consecutive base-pair planes to 6.6 Å [44b]. This process is certainly not disfavoured as to ΔG-figures: at saturation, one ethidium bromide molecule is intercalated per two base-pairs [44c]. As discussed in section 3.3.8 (*cf.* Table 3.12), a feeble preference of Trp for guanine, as compared to adenine, seems to exist and a marked preference for purine over pyrimidine bases (overlap area rule). These preferences may be strengthened if the corresponding cycles are fixed in adequate relative positions. The maximal overlap area rule together with superposition of opposite charges (*cf.* Figure 3.2, Chap. 3 and Figure 7.2a) suggest that the indolic residue of Trp may prefer to intercalate between two superposed guanine cycles and the phenolic residue of Tyr, between two thymidine cycles. Specificity in repressor-operator interactions may appear by spatial juxtapositions of aromatic residues of the repressor with regions of corresponding base composition on the operator, as illustrated in Figure 7.3.

A model for the binding of lac repressor to the operator was issued by Müller-Hill *et al.* [76]. The amino-acidic sequence of the lac repressor is partially sequenced [76, 77] and 50 *N*-terminal amino acids (which contain 4Tyr and 1His residue) are required for the repressor-operator interaction. Some informations about the lac operator sequence are also available (*cf.* section 7.4.1). The model of Müller-Hill *et al.* is a steric model, in which the (supposed) α-helix of the 50 *N*-terminal amino acid of repressor contacts, in the wide groove, the operator double helix. Specificity of binding is ensured by hydrogen bonds of the Tyr-17, Glu-18, Ser-21, Asp-25 and Glu-26 residues with functional groups at the edges of base-pairs and also Arg-22, His-29 and Lys-33 side chains are bound to phosphate residues and also to base-pairs. Nevertheless, while mutations producing Pro/Ser-16, Ala/Thr-19 and Val/Ala-53 substitutions annihilate *in vivo* repression, substitution of Glu-26 by either Leu, Tyr or Ser do not affect repression, which casts some doubts upon the model of Müller-Hill [78].

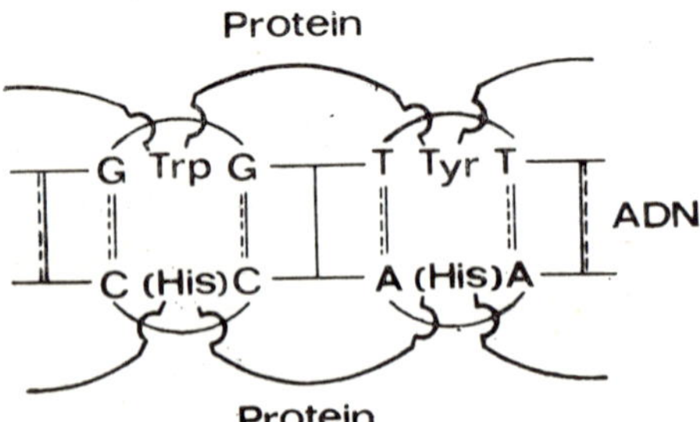

Figure **7.3** Hypothetic structure for specific repressor-operator interaction.

7.5. Specificity in immune phenomena

The immunological defence system involves some conditions regarding the specificity of the antigen-antibody reaction and — especially — the non-recognition of self [52]. Regardless of the mechanism of cell or antibody immune response and of which way lack of response against self is accomplished, the immunocytes, which detect (and elicite a response to) a non-self antigenic determinant should not detect those N self antigenic determinants with which they may come into contact. Each of these immunocytes seem to possess on their surface a species of antibody-combining site which plays the rôle of "recognising", A_0 and detects an alien antigenic determinant, B_0 at concentrations c_0 which are sometimes very low; however they do not recognise the self-antigenic determinants, B_i, at, sometimes very high, concentrations c_i.

A simple calculation may illustrate the specificity of this system. The antigenic determinants consist of not too large oligomers. The haploid complement of mammalian genome contains *ca.* 2.5×10^9 base-pairs; the overall length of proteins that can be coded is near 1×10^9 amino-acidic residues. The number of short oligopeptidic sequences, with overlapping, is about the same, and, in addition, there are polysaccharides and, eventually, nucleic acids. Thus, the number of self-antigenic determinants may exceed $N = 1 \times 10^9$. However it is possible that part of them cannot come into contact with antibodies due to structural reasons (internal arrangement of proteins), to the non-expression of the respective genes, to membraneous protection of the corresponding molecules, and, at the same time, mammalian genomes seem to present a high degree of repetativeness. It is, nevertheless, improbable that the number of "actual" self-antigenic determinants is smaller than $N = 10^7$.

7.5.1. *The antigens.*

The antigens are "non-self" polypeptides and polysaccharides; nucleic acids quite scarcely elicit antibodies (especially those of pro- and eukaryotes, containing no unusual bases) unless they have been complexed with proteins [53] or, DNA if it is at least partially denaturated (despiralised) [54]. One can elicit antibodies to a great number of micromolecules (of dye-type) if they are chemically bonded to polypeptides.

A prerequisite of antigenicity is a sufficiently high molecular weight. The minimum limit is *ca.* 4000—10 000 dalton, yet, seemingly, there exist some exceptions, such as angiotensin = 1031 dalton, and arsanyl-N acetyl-tyrosine = 451 dalton [55]. There are some proteins with very poor antigenicity: gelatin, poly-Ala, poly-Lys, poly-Glu. By adding to them aromatic amino acids (attachment of poly-L-Tyr, poly-L-Trp, poly-L-Phe or cyclohexyl Ala to ω-amino groups) these proteins and polypeptides may acquire a high antigenicity. Also, the poly-(Lys, Glu)copolymers are antigenic, unlike the corresponding homopolymers [55]. In case of DNA having antigenic properties, the antigenic portions are T-rich and, probably, denatured [54].

The size of antigenic determinant can be established by studying the inhibition of polymer precipitation with the corresponding antibody elicited against the oligomer covalently bonded to a protein, as a function of the number n of monomer units of the added oligomer. The antigenic determinant of some polysaccharides seems to consist of six monosaccharide residues ([53], Chap. 6); the smallest antigenic determinant known appears to be tri-alanine (from $^{+}H_3NAla_3 = Ala_{n-3}$-Carrier) [56]; for bradykinin coupled to polypeptide, nine amino acids; for the bradykininic sequence in akininogen, six amino acids [57]. the size of antibody combining sites (of about the same size or a little larger than the antigenic determinants) has been evaluated to an average of 10—12 amino acids by Burnet [52] and to 10—20 amino acids by Karush [58]. However, these experiments only reveal the haptenic aliquot from the antigenic determinant. There are experiments which show that the mouse spleen cells "primed" (as the first step towards eliciting antibodies) against a hapten (3.5-dinitro-4-hydroxy-phenylacetate) coupled to hen γ-globulin can be stimulated *in vitro* to a marked secondary response by hen γ-globuline [61]. That is, these cells have "receptors" for both the hapten and the carrier protein (for a certain part thereof). Also, while gelatin does not elicit antibody production, when it is coupled with 2% tyrosine, it does elicit antibodies to both poly-Tyr and gelatin [55]. The antibody combining site of the corresponding immunocytes probably interact with both hapten and a portion of the carrier protein.

The absence of response to self-antigenic determinants and the tendency towards responding more or less intensely to certain non-self antigenic determinants seem to be — at least to a very large extent — determined by the genetic particularities of the individual [59]. The mechanism of antibody elicitation is not fully elucidated. Anyway, it seems that each antibody-synthesising lymphocyte, with a few exceptions, synthesises a single type of antibody. The immunological defense is based either on antigen-serum antibody reactions, the youngest type phylogenetically and the most specific — or on a cellular defense by means of the specific intervention of another type of lymphocytes (which probably carry antibody-combining sites on their surfaces) [52]. The first stage or stages of elicitation comprise(s) the cleavage ("digestion") of antigens by macrophages and may be under the influence of a non-specific stimulus generated by organismic damage as a result of penetration of bacteria or, in general, of a foreign body. The macrophages probably are cells which present the antigenic determinant, in a more active form (in higher concentration) to some "initial" cells of the B- (marrow-derived) — or T- (thymus-dependent) series, that, in turn, "recognise" certain non-self antigenic determinants. These cells generate, after several stages, the elicitation of an antibody or cellular response to these foreign antigenic determinants [52, 60].

7.5.2. *Elicitation of antibodies and calculation of the size of antigenic determinants*

The process of recognition or non-recognition of a foreign (concentration c_0) or self (concentration c_i) antigenic determinant has been supposed to be a process of precipitation (or non-precipitation) of the agent on the surface of the immunocyte with the corresponding antibody sites [61]. Thus, we would be in the case described by equations (6.4), section 6.1. The affinity — G_{0i} for self-antigens (B_i) should thus be decreased, against to — G_{00} for the recognised antigen (B_0) by an amount ΔG_i, which may be obtained by the relations:

$$\frac{[A_0] \cdot c_0}{[A_0] \cdot c_i} > e^{-\frac{G_{00}-G_{0i}}{RT}} \tag{7.4}$$

$$\Delta G_i = G_{0i} - G_{00} > 2{,}3RT \log \frac{c_i}{c_0}; \quad i = 1, 2, \ldots, N \tag{7.5}$$

The "initial" recognition, as precipitation, involves a strong co-operative character for the antigen-recogniser immunocyte interaction; a second antigen molecule would be more strongly bound than the first, the third more than the second, etc., a sort of antigenic "crystallisation" stimulated by interaction with the surface of the recognising immunocyte. In favour of this hypothesis is the fact that in order to elicit antibodies, the antigen should be of macromolecular size, and the antibody-antigenic determinant affinities, as measured from equilibria in the solution, should yield smaller values than those resulting from the very low antigenic concentration (with the respective determinant), which are still able to elicit antibody production. The difference of these affinities is of 3—4 kcal mol^{-1} of antigenic determinant [62].

As regards the minimum concentrations of antigen able to elicit an antibody response, a discussion of data available in literature [61] indicates $c_0 \cong 10^{-6}$ M for "weak" antigens and $c_0 \geqslant 10^{-10}$ M for "strong" antigens. It is even more difficult to evaluate the maximum concentrations c_i, for each of the self-antigens with which the immunocytes come into contact. The serum globulins, with over half of their amino acid sequences identical, have concentrations of 10^{-3}—10^{-4} M; the antigenic determinants of blood vessel surfaces might have even higher concentrations. Considering a mean value, $c_i = 10^{-3}$ M, equation (7.5) yields $\Delta G = 4$ kcal mol^{-1} for "weak" antigens, and $\Delta G = 10$ kcal mol^{-1} for "strong" antigens. Most antigens consist of proteins, so that the monomers are the 20 amino acids—$\alpha = 20$. The importance of types of intermolecular forces in peptide-peptide interactions has been discussed in section 6.3, where it was shown that the substitution of an amino acid for another one within the antigenic determinant lowers the affinity towards the corresponding antibody by an average of $\Delta g = 2.0$ kcal mol^{-1}.

According to equation (6.18), section 6.4.3, with $N = 10^7$—10^9, $\alpha = 20$ for "weak" antigens; taking $\bar{\nu} = \Delta G/\Delta g = 2$, the antigenic determinant will be of $n = 7$—9 monomers, while for "strong" antigens ($\bar{\nu} = \Delta G/\Delta g = 5$) one obtains $n = 12 - 14$. This range ($n = 7$—14) is in agreement with what we know of the size of antigenic determinants (*cf.* section 7.5.1) except for Schaechter's [56] trialanine determinant. In the latter case, either the determinant also contains a part of the carrier or for some other reason, the immunocytes do not touch the oligoalanine sequences from the self-protein.

The poor antigenicity of nucleic acids may be accounted for as follows: DNA molecules are, within the conditions in the organism, double helical; RNA molecules have more complex spatial structure. An antibody against a foreign DNA or RNA oligonucleotidic sequence should be able to distinguish this sequence from among the $N \cong 10^9$ short oligonucleotide sequences corresponding to the self-genome (10^9, since one considers the high degree of genomic repetativeness). With this case, we only have four types of monomers, thus $\alpha = 4$. For protein-nucleic acid interactions $\Delta g = 2.1$ kcal mol^{-1} (section 7.4.1), and thus, by considering DNA or RNA as "weak" antigens ($\Delta G = 4$ kcal mol^{-1}) the foreign sequence of the length n should have to differ with at least $\nu = 2$ base-substitutions from all N self-sequences of the length n. Thus ($N = 10^9$, $\nu = 2$, $\alpha = 4$; equation (6.18) section 6.4.3) the antibody capable of such a performance, should have a combining site able to comprise $n \geqslant 19$ nucleotide residues (or their bases). Thus the organism must develop either antibodies of large combining sites (average molecular weight about 300 dalton, a nucleotide residue being 2.5 times larger than the amino acid) or no antibodies to nucleic acids so that the attack upon self-nucleic acids should not occur.

In case of viral nucleic acids the situation is more favourable, in the sense that they contain odd bases, which bring about the decrease of the size of antibody site required for distinguishing among self-nucleic acids. Let there be a viral antigenic determinant containing, 75% native bases, corresponding to a relative content $f = 0.25$ in self nucleic acids, and 25% unusual bases corresponding to a relative content $f = 10^{-3}$, self nucleic acids. According to relation (6.22) (section 6.4.3), we may use an affective $\bar{\alpha}$:

$$\alpha = (0.25^{0.75n} \times 10^{-3 \times 0.25n})^{-1/n} \cong 15. \qquad (7.6)$$

Equation (6.18) (section 6.4.3), with the other values as above ($N = 10^9$, $\nu = 2$, $\alpha = 15$) yields now $n \cong 10$, which is not too remote from the upper limits permitted for antibody sites (*cf.* section 7.5.1), even though a nucleotidic residue has a volume of twice that of an average amino acid.

Finally, one may not *a priori* exclude a mechanism by means of which the recognition is in two steps (*cf.* section 6.5). For example, let the macrophage "present" to immunocytes, in addition to foreign antigenic determinants, $\sqrt{N}$ self-antigenic determinants existing in higher concentrations — that is a decrease in ΔG, and of $\bar{\nu}$. In their turn, the immunocytes (more exactly, their antibody sites) should have to distinguish the "non-self" only from among $\sqrt{N}$ self-antigenic determinants, that — according to formula (6.18), section 6.4.3 would reduce n by a little less than 50%.

7.5.3. *Anti-aromatic nitroderivative myelomatous proteins*

Myelome is a cancer, of lymphoblastic system in which an antibody-producing cell undergoes uncontrolled proliferation and large quantities of a single antibody type are produced. Myelomes are thus allowing studies on well defined species of antibodies, including amino acid sequencing, structure and on interaction with antigenes, if molecules with high affinity for these antibodies are found.

Myelome proteins which, by chance, have a high affinity for dinitrophenyl derivatives, are those produced in cultures of two mouse plasmacytomes, termed MOPC-460 and MOPC-315 [90—93]. The 460 IgA myelome protein has two heavy chains (56 000 dalton) and two light chains (23 000 dalton). Its affinity towards a series of molecules (dissociation constants, *K*, 4 °C), according to Eisen [90], are listed in Table 7.4. As

TABLE 7.4 **Dissociation constant K(M) of various molecules in interaction with MOPC-460 protein**

Molecule	*K(M)*
ε-dinitrophenyl-L-Lys	2.5×10^{-5}
2,4-dinitronaphthol	3.9×10^{-4}
2,4,6-trinitrophenylaminocaproate	8.2×10^{-5}
menadione	3.6×10^{-4}
2,6-dinitrophenylaminocaproate	8.9×10^{-5}
4-nitrophenylaminocaproate	9.1×10^{-5}
purine-6-oilaminocaproate	1.1×10^{-2}
5-acetouracylcaproate	$> 10^{-3}$
ε-*p*-I-phenyl-sulphonyl-L-Lys	$> 10^{-3}$
ε-dimethylaminonaphthalene-subphonyl-L-Lys	$> 10^{-3}$
ε-*p*-toluene-sulphonyl-L-Lys	$> 10^{-3}$
1,8-anilinonaphthalene-sulphonate	$> 10^{-3}$

seen, this protein binds molecules of different shape and chemical nature such as menadione and nitro-derivatives of benzene and naphthalene (Figure 7.4) which share an aromatic planar ring and, especially electron-attracting substituents, conferring on them an electron-accepting character in charge-transfer—complex formation. The MOPC-315 protein has similar binding affinities towards dinitrophenyl derivatives [91, 97]. The hypothesis of Rosenstein *et al.* [92] of two nearby competing sites in the

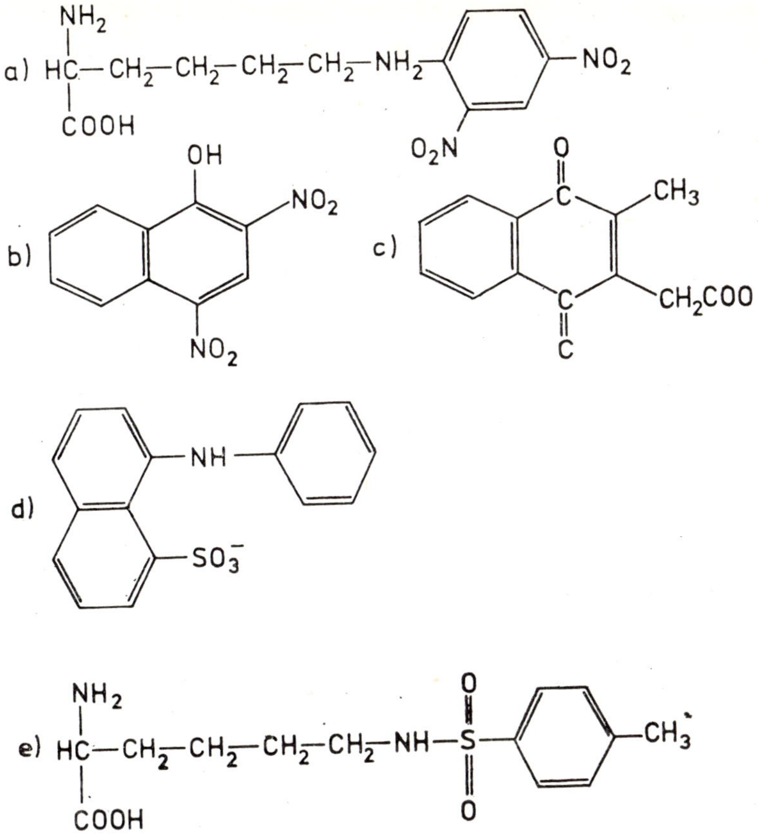

Figure 7.4 Molecules which bind to the MOPC-460 myeloma protein: (a) 2,4 dinitrophenyl-lysine; (b) 2.4 dinitronaphthol; (c) menadione; (d) 1,8 anilinonaphthalene-sulphonat (does not bind); (e) ε-toluen-sulphonyllysine (does not bind).

MOPC-460 protein, one for menadione, the other for dinitrophenyl, does not seem plausible. Rather, the MOPC-460 binding site should have an electron donating character for charge-transfer complexes, and, in reality, affinity label studies [91] and circular dichroism studies [97] strongly suggest the presence in the binding sites of these proteins of tryptophane, the only amino acid with electron-donor character for charge-transfer complexes (*cf.* section 3.3.9).

Kinetic studies on binding and dissociation from the MOPC-315 protein were also performed [93]. Thus, the binding rate for DNP-Lys to this protein (25 °C) is $1.3 \times 10^8 M^{-1}s^{-1}$, near the diffusion limit, the dissociation rate 72 s^{-1}, the resulting binding constant 1.8×10^6 M^{-1}. Positive charge on the molecule stimulates binding: for DNP-Gly, the dissociation rate is 900 s^{-1}, the binding constant $3.0 \times 10^5 M^{-1}$ [93a].

Antibodies elicited against polynitro-derivatives in various animals were also studied and compared to the MOPC-315 and 460 proteins [94—96]. It is interesting to note that Raman-resonance spectroscopic studies on complexes between such an antibody (this time heterogeneous, elicited in rabbits) and azo-dyes containing the dinitrophenyl group indicate that charge-transfer complex formation does not seem to be implied in this case [96]. The binding constant of rabbit antibody for DNP-Lys is also about 10^7 M^{-1}, with an exothermal binding enthalpy, $\Delta H \cong -15$ kcal mol^{-1} (25 °C [95]). This may suggest, the quite natural conclusion, that the elicited antibodies are not necessarily the proteins with the highest theoretical possible affinity for the antigen, but only those proteins the with highest affinity for the antigen among the potential antibodies available in the immunity defence system of the animal.

7.5.4. *The antigen-antibody neutralisation reaction*

This may be considered, by overtly simplifying, as a precipitation reaction of the type [61]:

$$[C]\,[D_1]\,...\,[D_\mu] \geqslant P' = e^{-\frac{G'}{RT}}, \tag{7.7}$$

where C is an antigen, and $D_1 ... D_\mu$ are antibodies against μ determinants on the antigenic surface. Here μ would be the minimum number of antigenic determinants which should interact with antibodies for the neutralisation reaction to occur.

The affinity $-G'$ can be connected to the affinities of antibodies D_1, ..., D_μ to antigenic determinants on C, that can be deduced from studies on equilibria in solution [53]. If we denote these affinities towards antigenic determinants by $-g_i'$, the simplest relation between $-G'$ and g' would be:

$$G' = \sum_{i=1}^{\mu} g_i' = \mu \overline{g}', \tag{7.8}$$

and the condition for the production of a neutralisation reaction can be written:

$$[D] > e^{\overline{g}'/RT} \cdot [C]^{-1/\mu}. \tag{7.9}$$

With sufficiently high "valence", μ, for antigen, the factor $[C]^{1/\mu}$ will not decrease below 0.01, and thus for the $-g' = 5 \ldots 12$ kcal mol^{-1} values obtained from equilibrium studies, the antibody concentrations requested for the precipitation of antigen will be of $[D] \geqslant 10^{-2} \ldots 10^{-7}$ M. If the overall serum concentration of γ-globulins is considered as 10^{-4} M and accepting 10^{-3} % for the mean percentage of "natural" individual antibodies [53], the serum concentration of a "natural" antibody will be *ca.* 10^{-9} M. The quantity of serum "natural" individual antibody (i.e. that existing prior to elicitation of the production of the corresponding antibody as a result of a primary and/or secondary response) is therefore insufficient to neutralise the antigens (self or non-self of the complementary configuration) with which it comes into contact.

Equation (7.9) does still indicate that the higher the valence (i.e. the molecular weight) of an antigen, the lower the antibody concentration required for its precipitation [61]. For example, for a certain amount of antigen with $\mu = 3$, corresponding to $[C] = 10^{-6}$ M, the factor $[C]^{-1/\mu}$ from equation (7.9) has the value of 100. For the same amount of an antigen with $\mu = 30$, i.e. with a molecular weight by about $10^{3/2}$ times higher, one would get $[C] = 10^{-7.5}$ M and $[C]^{-1/\mu} \cong 2$ only. This inference of equation (7.9) does agree to the general rule, that the higher the molecular weight of antigen, the smaller the amount of antibody from precipitate [63] which suggest that the higher the molecular weight of the antigen, the smaller the minimum amount of antibody required for precipitation of antigen.

Finally, one of the reasons for the generally poor efficiency of immunological defence in cancer might also be the following. For the precipitation reaction or for favouring by the antibody of the phagocyte reaction of a "foreign" cell, the hypothesis that led to relation (7.7) suggests the necessity that antibodies exist against a minimum number μ of species of antigenic determinants on the surface of this cell. The malignant cells, originating in the normal cells of organism, might present a too small (smaller than μ) number of species of determinants differing from those on surface of normal cells. It is these distinct determinants that can be recognised as "non-self", and only against these can antibodies be elicited. In these conditions malignant cells would elicite too few antibody species to satisfy inequality (7.7).

7.6. Some considerations concerning drug design

The search for drugs to cure different diseases is as old as medicine and large scale systematic work in our century was very successful, although its theoretical base may have been rather weak. Ehrlich's Salvarsan (Figure 7.5) was the result of a search on some hundreds of substances, starting probably from the observation of specific staining of bacteria by some azo-dyes and the introduction of arsenic to give toxicity. Other examples are listed in Albert's "*Selective Toxicity*" [64]. One may give also examples of very cumbersome but not very successful research on such lines — for example the hundreds of thousands of compounds listed in the "*Cancer Screening Data*", and the not very successful results of cancer chemotherapy [65].

Principles for guiding the search for structures with a selected biological activity are thus badly needed, and the work of Cammarata, Hansch and other authors mentioned already (sections 3.1.3, 3.3.7, 4.4.3) is on these lines. In these works correlational equations are set up which give the biological activity as a linear function of Hammett σ-constant and hydrophobicity π-constants, for a given class of compounds; quadratic terms are also sometimes included (for example equations (4.14)—(4.15), section 4.4.3). The sign and size of the optimised coefficients indicate what sort of substitution (a more hydrophobic group, more electron-attracting group, etc.) should favour the given type of biological activity. If second-power terms are included, one can indicate also optimum values for the corresponding structural parameters (optimum σ, π-values, etc.) [66]. Unfortunately such correlational equations usually encompass narrow ranges of chemical structures, within which the linear correlation coefficient often attains values above $r = 0.95$. Attempts to correlate biological activity with several structural intermolecular force parameters (see section 6.3) allowed to encompass quite large ranges of chemical structures but yields low r values ($r = 0.7$). A set of parameters which permit

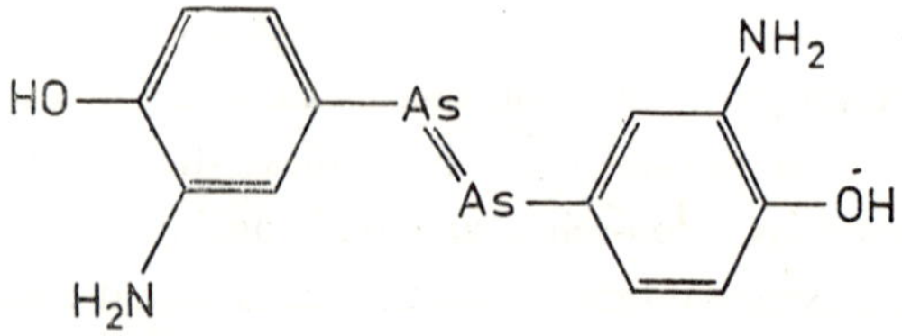

Figure 7.5 The Salvarsan molecule.

structure-activity correlations both with high *r* values and within large ranges of chemical structures would be most useful for drug design.

Another type of search-strategy stems from the model of Free and Wilson [66, 67]. In this model, the biological activity is considered to be the sum of contributions from molecular segments which may contain different substituents. On the other hand for a certain minimum number of biological activities one calculates the effects of substituents in different segments and one can predict, within the given set of segments and substituents, which molecule should be the most active one. The Free and Wilson model is also difficult to generalise so as to cover a wider range of chemical structures.

For several types of biological activities data are available for a great number of compounds, often belonging to various types of chemical structures. The amount of calculations required to use such large series of data in forecasting the optimal chemical structure may impose a logico-analytical scheme and computerisation. Such search strategies were proposed by Hiller *et al.* [68] based either on a perceptronic, a logico-structural, or a topographic algorithm. Some tests on different sets of chemical compounds gave an average realiability of 75—80% in predicting active or non-active compounds respectively. A key problem of such computerised search methods is to found an adequate "dictionary" of structural parameters for characterising the compounds.

A principle which does not seem to be widely recognised is the need of a minimal molecular size for sufficiently high specificity. Drugs such as sulphamides block some enzymes of bacteria which differ sensibly from the corresponding ones of mammalian cells ([64], Chap. 3); even if these drugs combine with several proteins (not key enzymes!) of the mammalian cells this may have no dramatic biological effects. High requirements of specificity impose the use of molecules of increasing size and complexity — the antibiotics of the last decade for example. Cancer, as suggested especially by the viral theories, resides in some defect within the genome, perhaps in some viral oncogenes incorporated in the host genome. The highest efficiency of a drug should be obtained if it acts on the first "amplifying" level of cell regulation, on the translation of these oncogenes. Therefore, the ideal anticancer drugs should have a specificity comparable to repressor proteins and it is rather hopeless to search for some simple molecule which should be highly effective in some cancer [69]. But with macromolecules the number of possible structural combinations reaches astronomic figures and it would be very difficult to find a strategy of search in this case.

One may search for an efficient anticancer therapy in the field of immunological defence. In most cases there seems to exist an immune defence against the development of the tumour, but not a sufficiently efficient one; sometimes sera of cancer patients contains inefficient antibodies which even protect cancer cells against phagocytosis [70]. Also, one should not forget that there are microbial diseases in which, even after a prolonged chronic period, sufficient for complete mobilisation of the immunity defence of the organism, there is a lethal end in the absence of efficient antibiotica — tuberculosis for example.

One idea to use immunological defence in cancer is the one quoted in 1969 by Zilber and Abelev [71]: to make tolerant against the patient's normal tissue to a group of newborn animals of short lifespan, and, after the animals aquire immunological maturity, to inject them with a preparation of tumour cells, with the hope they will elicit antibodies only against tumour antigenic determinants. This antiserum could then be used to cure the respective cancer. Since then, although a couple of years have passed, this idea does not seem to have acquired therapeutic success.

Two possible causes may determine the low efficiency of immunological response in cancers (possibly both):

(i) The antigenic determinants (which make tumoral cell surfaces differ from those of the corresponding normal type of cells) may have rather small structural differences from the normal ones and thus have a rather weak "non-self", i.e. antibody eliciting, character.

(ii) The specific tumoral determinants have a too low density on the cell surface and even in the presence of antitumor antibodies cannot start the subsequent reactions which distroy the "recognised" cell. This implies the assumption that a minimal antibody density on the cell surface is required in order that the cell is destroyed, eventually in the sense of the solubility product — precipitation relation described by equation (7.7), section 7.5.4.

The second cause suggests a potentiation of the immunological anticancer defence by injection of a serum containing antibodies against "normal" antigenic determinants of the respective cancer cells. The quantity of the "anti-normal" serum should be balanced such as to satisfy, together with the "anticancer" antibodies, the solubility product relation (equation (7.7), section 7.5.3) against the cancer cells but not also the one against the corresponding normal cells.

One may, nevertheless, ask if the scarce success of classic cancer therapy is not due to the inadequacy of screening procedures. The main limitation of classic anticancer drugs is that they are effective against rapid proliferating cells, both cancerous and normal (especially the haematopoietic system) [65]. The screening and tests of the drugs seem to be performed on a standard set of experimental tumours which indicates especially the antiproliferative action of the drug, not also its selective permeability with respect to the tumour cells of the patient. There seem to be sensible differences between the cell surfaces of normal and malignant cells [72] and one should also expect differences in permeabilities. Such permeability differences may explain differential *(in vitro)* effects of one drug on related lines of normal or malignant cells, rather than differences in proliferation rates [73]. One should perhaps test several drugs, of a given class, with sufficient general antiproliferative action, on their relative permeabilities in haematopoietic stem cells *versus* cells of the interesting human tumour (freshly extracted, in a medium as near as possible to the *in vivo* situation) [74]. Such tests, coupled with correlational techniques described previously, may provide a more efficient design of drugs against specific human tumours.

REFERENCES

1. J. L. Webb, *Enzyme and Metabolic Inhibitors*, Academic Press, New York (1963) Ch. 16.
2. J. W. Drake, *Nature*, **221**, 1132 (1969).
3. C. Bresch and R. Hausmann, *Classical and Molecular Genetics* (in German), Springer, Berlin, (1970) Ch. 6.
4. J. H. Knowles, *J. Theoret. Biol.*, **9**, 213 (1965).
5. (a) G. L. Neil, C. Niemann and G. E. Hein, *Nature*, **210**, 903 (1966); (b) L. M. Vornick and J. S. Fruton, *Proc. Nat. Acad. Sci. U.S.A.*, **68**, 257, (1971).
6. R. Lowrien and T. Linn, *Biochemistry*, **6**, 2281 (1967).
7. B. H. H. Hofstee, *Nature*, **213**, 42 (1967).
8. Z. Simon, *J. Theoret. Biol.*, **9**, 414 (1965).
9. C. T. Caskey, *Quart. Rev. Biophys.*, 3, 295 (1970).
10. H. E. Kubitscheck and T. R. Henderson, *Proc. Nat. Acad. Sci. U.S.A.*, **55**, 512 (1966).
11. M. Nirenberg, *Angew. Chemie*, **87**, 1017 (1969).
12. G. von Ehrenstein, *Cold Spring Harbor Symp. Quant. Biol.*, **31**, 705 (1966).
13. M. Chamberlin and P. Berg, *J. Molec. Biol.*, **8**, 708 (1964).

14. (a) J. B. Weinstein, S. M. Friedman and M. Ochoa Jr., *Cold Spring Harbor Symp. Quant. Biol.*, **31**, 671 (1966); (b) V. A. Gynewskaia, I. V. Skarlat, N. O. Kalinina and V. I. Agol, *Molek. Biol.*, **5**, 132 (1971); (c) L. B. Loftfield, *Biochem. J.*, **89**, 82 (1963).
15. (a) J. Davies, *Cold Spring Harbor Symp. Quant. Biol.*, **31**, 665 (1966); (b) S. Perzynskii, P. Chomezynski and P. Szafranski, *Biochem. J.*, **114**, 437 (1969); (c) S. M. Friedman, R. Berezeney and I. B. Weinstein, *J. Biol. Chem.*, **243**, 5044 (1968).
16. (a) R. B. Loftfield and E. A. Eigner, *J. Biol. Chem.*, **240**, 1482 (1965); (b) H. Calendar and P. Berg, *J. Molec. Biol.*, **26**, 39 (1967).
17. P. Claverie, *J. Molec. Biol.*, **56**, 75 (1971).
18. J. Miklos, *Studia Biophys.*, **28**, 223 (1972).
19. C. R. Woese, *J. Theoret. Biol.*, **26**, 83 (1970).
20. V. I. Danilov, K. B. Tolpygo and O. V. Shramko, *Tsitol. Genet., Akad. Nauk Ukr. S.S.R. (Cytology & Genetics, Ukr. Acad. Sci.)*, **2**, 195 (1968).
21. F. H.C. Crick, *J. Molec. Biol.*, **19**, 548 (1966).
22. J. L. Nichols, *Nature*, **225**, 147 (1970); *cf.* also *Nature*, **225**, 127 (1970) and **230**, 206 (1971).
23. (a) *Nature*, **228**, 608 (1970); (b) *Nature*, **233**, 526 (1971); (c) G. Sundharadas, J. R. Katze, D. Söll, W. Königsberger and P. Lengyel, *Proc. Nat. Acad. Sci. U.S.A.*, **61**, 693 (1968); (d) A. D. Mirazbekov, L. Ia. Kazarinova and A. A. Baev, *Molek. Biol.*, **3** 879, 909 (1969); (e) G. Högenauer, *Europ. J. Biochem.*, **12**, 527 (1970); (f) P. M. Bhargayn, T. Pallaick and E. Premkumar, *J. Theoret. Biol.*, **29**, 447 (1970); (g) G. N. Ramachandran and A. V. Lakshminarayanan in *Macromol. Storage and Transfer Biol. Inform.* Proc. Symp. Bombay, (1969) pp. 17—20.
24. Eisenger, Feuer and Yamane, cited in *Nature*, **231**, 420 (1971.).
25. C. A. Thomas, Jr., *Progr. Nucleic Acid. Res. Molec. Biol.*, **5**, 315 (1966).
26. (a) B. L. Conaughy and B. J. McCarthy, *Biochim. Biophys. Acta*, **149**, 180 (1967); (b) Idem, *Biochemical Genetics*, **4**, 409 (1970).
27. S. K. Niyogi, *J. Biol. Chem.*, **244**, 1576 (1969).
28. P. M. B. Walker, *Progr. Nucleic Acid Res. Molec. Biol.*, **4**, 301 (1969).
29. (a) J. O. Bishop, *Biochem. J.*, **113**, 805 (1969); (b) M. L. Bernstiel, B. H. Sells and J. F. Purdon, *J. Molec. Biol.*, **63**, 21 (1972).
30. W. Gilbert and B. Müller-Hill, *Proc. Nat. Acad. Sci. U.S.A.*, **57**, 1891, 2415 (1967).
31. M. Ptashne, *Proc. Nat. Acad. Sci. U.S.A.*, **57**, 306 (1967); *Nature*, **214**, 232 (1967).
32. S. Udaka, *Nature*, **228**, 336 (1970).
33. J. S. Parks, M. Gottesman, K. Shimada, R. A. Weisberg, R. L. Periman and I. Pastan, *Proc. Nat. Acad. Sci. U.S.A.*, **68**, 1891 (1971).
34. G. Wilcox, K. J. Clementson, D. V. Santi and E. Engelsberg, *Proc. Nat. Acad. Sci. U.S.A.*, **68**, 2145 (1971).
35. A. D. Riggs, G. Reiness and G. Zubay, *Proc. Nat. Acad. Sci. U.S.A.*, **68**, 1222 (1971).
36. A. A. Travers, *Nature*, **223**, 1107 (1969).
37. S. D. Barbour, C. Cross and A. Novick, *J. Molec. Biol.*, **33**, 967 (1968).

38. B. Müller-Hill, *Angew. Chem.*, **83**, 195 (1971).
39. (a) T. F. Smith and J. R. Sadler, *J. Molec. Biol.*, **59**, 273 (1971); (b) J. R. Sadler and T. F. Smith, *idem*, **62**, 139 (1971).
40. (a) A. D. Riggs, H. Suzuki and S. Bourgeois, *J. Molec. Biol.*, **48**, 67 (1970); (b) A. D. Riggs, R. F. Newby and S. Bourgeois, *idem*, **51**, 303 (1970); (c) A. D. Riggs, S. Bourgeois and M. Cohn, *idem*, **53**, 401 (1970).
41. A. Gierer, *Nature*, **212**, 1480 (1966).
42. S.-Y. Lin and A. D. Riggs, *Nature*, **228**, 1184 (1970).
43. Z. Simon, *Studia Biophys.*, **26**, 179 (1971).
44. (a) L. S. Lerman, *Proc. Nat. Acad. Sci. U.S.A.*, **49**, 94 (1963); (b) W. Fuller and M. J. Warring, *Ber. Bunsenges. Phys. Chem.*, **68**, 805 (1964); (c) J. Paoletti and J. B. Le Pecq, *J. Molec. Biol.*, **59**, 43 (1971).
45. H. Busch, W. J. Steele, L. S. Hnilica, Ch. W. Taylor and H. Mavioglu, *J. Cell. Comp. Physiol.*, **62**, 95 (1963).
46. J. Bonner, M. E. Dahmus, D. Fambrough, R. C. Huang, K. Marushige and D. Y. H. Tuan, *Science*, **159**, 47 (1968).
47. M. Artman and J. S. Roth, *J. Molec. Biol.*, **60**, 291 (1971).
48. J. T. Shapiro, M. Leng and G. Felsenfeld, *Biochemistry*, **8**, 3219 (1969).
49. M. Leng and G. Felsenfeld, *Proc. Nat. Acad. Sci. U.S.A.*, **56**, 1325 (1966).
50. Z. Simon, *Studia Biophys.*, **28**, 179 (1972).
51. J. T. Shapiro, M. Leng and G. Felsenfeld, *Biochemistry*, **8**, 3219 (1969).
52. M. F. Burnet, *Cellular Immunology*, Melbourne University Press, Melbourne, (1969).
53. E. L. Kabat, *Structural Concepts in Immunology and Immunochemistry*, Holt, Rinehard and Winston, New York, (1968).
54. D. B. Saprygin and A. M. Poverennyi, *Molek. Biol.*, **3**, 815, (1969).
55. M. Sela, *Naturwiss.*, **56**, 206 (1969).
56. I. Schaechter, *Nature*, **228**, 639 (1969).
57. J. Fischer, J. Spragg, R. C. Talamo, J. V. Pierce, K. Suzuki, K. F. Ansten and E. Haber, *Biochemistry*, **8**, 3750 (1969).
58. F. Karush, *Advan. Immunology*, **2**, 1 (1962).
59. (a) see M. Sela, *First National Symposium on Immunology* (Abstracts), Bucharest, (1971) p. 66; (b) A. Sulica, E. Mozes, G. Shaerer and M. Sela, *idem*, p. 82; (c) *cf.* the Section on Molecular Genetics of Antibody Formation, in: *Abstracts of 7th FEBS-Meeting*, Varna, (1971) pp. 89—98.
60. V. Gheţie, *Antibodies in Action* (in Romanian), The Romanian Encyclopaedic Publishers, Bucharest, (1971).
61. Z. Simon and V. Gheţie, *Rev. Roum. Boichim.*, **8**, 261 (1971).
62. Data cited by S. Cohen, *J. Theoret. Biol.*, **27**, 19 (1970).
63. W. C. Boyd, *Fundamentals of Immunology*, Interscience Publ., New York, London, (1956) pp. 311—312.
64. A. Albert, *Selective Toxicity* (Russ. transl.), Izd. In. Lit. Moscow, 1953; John Wiley New York, (1951).

65. see for example L. F. Larionov, *Chemotherapy of Malignant Tumours* (Russian), Moscow (1962); E. A. Plattner, editor, *Antimetabolites and Chemotherapy*, Elsevier, Amsterdam, (1964); Papers of the *Symposium of European Societies of Anticancer Research*, Budapest, (1972).
66. for a review see I. Schwartz, *Studii Cercet. Chimie*, **21,** 721 (1973).
67. S. M. Free and I. N. Wilson, *J. Med. Chem.*, **7,** 398 (1964).
68. S. A. Hiller, A. B. Glaz, V. E. Holender, L. A. Rastrighyn and A. B. Rozenblit, *Khimikofarmatsevtichesky Zhurnal* (Russ.), **11,** 18 (1972).
69. Z. Simon, *Rev. Roum. Biochimie*, **6,** 239 (1969).
70. V. Ia. Shatz, *Tsitologhia* (Russ.), **14,** 3 (1972); J. Tooze, *Nature*, **233,** 28 (1971).
71. G. L. Abelev and L. A. Zilber, *Virology and Immunology of Cancer*, Academic Press, New York, London (1969); E. Levi and A. M. Schachtman, *Cancer Res.*, **23,** 1566 (1963).
72. see for example M. M. Burger and K. D. Noonan, *Nature*, **228,** 512 (1970); W. R. Loewenstein and Y. Kanno, *J. Cell. Biol.*, **33,** 225 (1967); R. A. Good, *Proc. Nat. Acad. Sci. U.S.A.*, **69,** 1026 (1972).
73. M. M. Blacket and R. E. Millard, *Nature*, **244,** 300 (1973).
74. M. Ciuştea — personal communication.
75. P. H. von Hippel and J. D. Mc Ghee, *Ann. Rev. Biochem.*, **41,** 232, (1972).
76. K. Adler, K. Beyereuther, E. Fanning, N. Geissler, B. Cronenborn, A. Klemm, B. Müller-Hill, M. Pfahl and A. Schmitz, *Nature*, **237,** 322 (1972).
77. T. Platt, J. C. Files and K. Weber, *J. Biol. Chem.*, **248,** 110 (1973).
78. K. Weber, T. Platt, D. Ganem and J. H. Miller, *Proc. Nat. Acad. Sci. U.S.A.* **69,** 3624 (1972).
79. C. F. Springgate and L. A. Loeb, *Proc. Nat. Acad. Sci. U.S.A.*, **70,** 245 (1973).
80. (a) Rose and Dudock, *Biochem. Biophys. Res. Commun.*, **49,** 339 (1972); (b) Crothers, Seno and Soll, *Proc. Nat. Acad. Sci. U.S.A.*, **69,** 3063 (1972); (c) Simsek and Ray Bhandary, *Biochem. Biophys. Res. Commun.*, **49,** 508 (1972); cited by *Nature*, **240,** 254 (1972).
81. J. Ninio, *Progr. Nucleic. Acid Res.*, **1973,** 13.
82. (a) R. Holliday and G. M. Tarrant, *Nature*, **238,** (1972); (b) R. B. Loftfield and D. Vanderyagt, *Biochem. J.*, **128,** 1353 (1972).
83. R. Rice and H. Fraenkel-Conrat, *Biochemistry*, **12,** 181 (1973).
84. (a) R. Mulivar and H. P. Rappaport, *J. Molec. Biol.*, **76,** (1973); (b) M. Jarus, *Biochemistry*, **11,** 2050, (1972).
85. Goldberg, *Proc. Nat. Acad. Sci. U.S.A.*, **69,** 422 (1971); cited by *Nature*, **236,** 143, 199 (1972).
86. L. E. Orgel, *Proc. Nat. Acad. Sci. U.S.A.*, **49,** 517 (1963); *Nature*, **243,** 441 (1973).
87. J. C. Arcos, R. S. Sohal, Sh. Ch. Sun, M. F. Argus and G. E. Burch, *Exp. and Molec. Pathol.*, **8,** 49 (1968).
88. A. Jobe and S. Bourgeois, *J. Molec. Biol.*, **72,** 139 (1972).
89. T. Maniatis and M. Ptashne, *Proc. Nat. Acad. Sci. U.S.A.*, **70,** 1531 (1973).
90. B. M. Jaffé, E. S. Simms and H. N. Eisen, *Biochemistry*, **10,** 1693 (1971).
91. J. Haimovich, H. N. Eisen, E. Hurwitz and D. Givol, *Biochemistry*, **11,** 2389 (1972).

92. R. W. Rosenstein, R. A. Musson, M. Y. K. Armstrong, W. H. Konigsberg and F. F. Richards, *Proc. Nat. Acad. Sci. U.S.A.*, **69**, 877 (1972).
93. (a) J. Pecht and D. Haselkorn, *Commun. I. Isr. Chem. Congress*; (b) D. Haselkorn, I. Pecht, S. Friedman, A. Yaron, D. Givol and M. Sela, *Isr. J. Chem.*, **9** (1971), 20; (c) I. Pecht, D. Givol and M. Sela, *J. Molec. Biol.*, **68**, 241 (1972).
94. G. S. Hsia and J. R. Little, *Biochemistry*, **10**, 3742 (1971).
95. B. G. Barisas, S. J. Singer and J. M. Sturtevant, *Biochemistry*, **11**, 2741 (1972).
96. P. R. Carey, A. Froese and H. Schneider, *Biochemistry*, **19**, 2198 (1973).
97. J. H. Rockey, P. C. Montgomery, B. J. Underdown and K. J. Dorrington, *Biochemistry*, **11**, 3172 (1972).
98. W. Gilbert and A. Maxam, *Proc. Nat. Acad. Sci. U.S.A.*, **70**, 3581 (1973).

APPENDIX 1

Calculation of the entropy of mixing of reactions in solution

Most of chemical reactions occurring at cellular level, take place in aqueous solutions. If the hydration of reactants and that of products is similar, the entropy of several reactions is mainly determined by the entropy mixing ΔS_{mix}, which for typical reactions is géven by:

$$A + \sigma S \rightarrow B + C + (\sigma - 1) S; \quad \Delta S'_{mix} \tag{A.1.a}$$

$$A + \sigma S \rightarrow B + C + \sigma S; \quad \Delta S''_{mix} \tag{A.1.b}$$

Shere is the solvent molecule, while σ is the number of moles of solvent from a litre of solution wherein the activity of the solutes is equal to unity, i.e. in the case of ideal solutions 1 mol l^{-1}. For molar solutions in water, the number of moles of H_2O l^{-1} of solutions, $\sigma \cong 50$, thus $\sigma \gg 1$.

In case of reaction (a), for 1 l of solution we have N_A molecules A and σN_A molecules of the initial solvent, and after reaction N_A molecules for each of A and B, and $(\sigma - 1) N_A$ molecules of solvent. From the usual formula for S:

$$\Delta S'_{mix} = S_{final} - S_{initial} = k \ln \frac{[N_A(\sigma + 1)]!}{N_A! N_A! [N_A(\sigma - 1)]!} -$$

$$- k \ln \frac{[N_A(\sigma + 1)]!}{N_A! (N_A\sigma)!} = k \ln \frac{(N_A\sigma)!}{N_A! [N_A(\sigma - 1)]!} =$$

$$= k \ln \frac{(N_A \sigma)^{\sigma N_A}}{N_A^{N_A} . [N_A(\sigma - 1)]^{N_A(\sigma - 1)}} .$$

Stirling's approximation was used: $X! = (x/e)^x$. Further on:

$$\Delta S'_{mix} = N_A k \, [\sigma \ln N_A\sigma - \ln N_A - (\sigma - 1) \ln N_A (\sigma - 1)] =$$

$$= R [\sigma \ln \sigma - (\sigma - 1) \ln (\sigma - 1)] = R \left[\ln \sigma + (\sigma - 1) \ln \frac{\sigma}{\sigma - 1} \right] =$$

$$= R \left[\ln \sigma + (\sigma - 1) \cdot \frac{1}{\sigma} \right] \cong R(1 + \ln \sigma).$$

(Therefere we took into account that $\sigma/(\sigma - 1) \cong 1 + 1/\sigma$, and since $1/\sigma \ll 1$, $\ln(1 + 1/\sigma) \cong 1/\sigma$).
With $R = 2$ cal mol^{-1} deg^{-1} and $\sigma = 50$, one obtains $\Delta S_{mix} = +9.8$ cal mol^{-1} deg^{-1}.

Similarly, for equation (A.1.b):

$$\Delta S''_{mix} = k \ln \frac{[N_A(\sigma + 2)]!}{(N_A\sigma)!\, N_A N_A!} - k \ln \frac{[N_A(\sigma + 1)]!}{N_A!(N_A\sigma)!} =$$

$$= k \ln \frac{N_A(\sigma + 2)!}{N_A!(N_A(\sigma + 1)!} = R[(\sigma + 2) \ln N_A(\sigma + 2) - \ln N_A - \tag{A.1.d}$$

$$- (\sigma + 1) \ln N_A(\sigma + 1)] \cong R[1 + \ln(\sigma + 2)].$$

For $\sigma = 50$, $\Delta S''_{mix} = +9.9$ cal mol^{-1} deg^{-1}. Thus, at 300 °K, the contribution $-T\Delta S$ of the variation of entropy of mixing to the free enthalpy, ΔG, of reaction is of $-2.9 \ldots 3.0$ kcal mol^{-1}.

APPENDIX 2

Calculation of the linear correlation coefficient r

Let us consider a series of compounds $i = 1, 2, \ldots, N$ and X_i, Y_i two parameters characterising these compounds, two properties of these compounds. One assumes that the parameter Y depends upon the parameter X, but also upon some other unknown parameters. A measure of the degree of dependence of parameter Y upon X is the linear correlation coefficient r, given by the equation:

$$r = \frac{\sum_{i=1}^{N} (X_i - \overline{X})(Y_i - \overline{Y})}{\sqrt{\sum_{i=1}^{N} (X_i - \overline{X})^2 \cdot \sum_{i=1}^{N} (Y_i - \overline{Y})^2}} \quad \therefore \overline{X} = \frac{1}{N} \sum_{i=1}^{N} X_i,$$

$$\overline{Y} = \frac{1}{N} \sum_{i=1}^{N} Y_i.$$

The formula is valid only for a sufficient large number N of compounds. If parameters X and Y are independent, $r = 0$; the maximal value, indicating a linear dependence between X and Y is $r = 1$.

APPENDIX 3

Definition of the "biological activity", A

The "biological activity", A_i^{exp} as usually used in structure-activity correlations is the logarithm of a quotient of two concentrations, c_i and c_0:

$$A_i^{exp} = \log c_i/c_0$$

with c_i the concentration of compound *i* which produces the same biological effects as the concentration c_0 of the standard compound. For the respective series for hormonal oligopeptides (reference [12] Chap. 6), as an example, the effect is a certain increase of the arterial pressure (or other physiological effect) produced by the concentration c_i of compound *i* or by the concentration c_0 of the standard compound. For haptene-antibody interactions the ratio c_i/c_0 is a measure of the relative affinity of compound *i*, as compared to the standard compound, towards the antibody elicited in an animal (rabbit) against the standard compound (haptene) coupled to a carrier protein. The standard compound coupled to the protein is injected repeatedly in to the animal, till a massive antibody response is obtained against this antigen. To an alliquots of this serum, a concentration c_0 of the labelled standard compound is added and also increasing concentrations of the unlabelled compound "*i*". The antibody, which binds variable quantities of the standard and the compound "*i*", is sedimented on a filter which retains colloidal particles. The concentration c_i for the compound "*i*" is the one which reduces to 50% the radioactivity of the antibody deposited on the filter.

The relation between the relative activity A_i^{exp} and the affinity decrease, $-\Delta G_i$, of compound "*i*", as compared to the standard, at usual temperatures (*ca.* 300 K) is given by:

$$\Delta G_i(\text{kcal mol}^{-1}) = 1.4 \log \frac{c_i}{c_0}$$

APPENDIX 4

The multiple correlation method

Let us consider a set of *N* compounds, $i = 1, 2, ..., N$, each with the activity A_i^{exp} and *m* structural characteristics of these compounds, described by the parameters σ_{ij}; $j = 1, 2, ..., m$. Generalising the linear Hammett type relations, a linear equation is found which correlates the m structural characteristics with the biological activity:

$$A_i^{calc} = \alpha + \sum_{j=1}^{m} \beta_j \sigma_{ij}.$$

The coefficients α and β_j are obtained by minimising the quadratic error Y with respect to these coefficients:

$$Y = \sum_{i=1}^{N} (A_i^{calc} - A_i^{exp})^2$$

$$\frac{\partial Y}{\partial \alpha} = 0 \qquad \frac{\partial Y}{\partial \beta_j} = 0 \qquad j = 1, 2, \ldots, m$$

By equating to zero the derivatives of Y with respect to the coefficients α, $\beta_1, \beta_2, \ldots, \beta_m$, a linear system of equations in α and β_j is obtained, whose solution represents the optimal values for these coefficients. This system of $m + 1$ equations is:

$$N\alpha + \sum_{j=1}^{m} \left(\sum_{i=1}^{N} \sigma_{ji} \right) \beta_j = \sum_{i=1}^{N} A_i^{exp},$$

$$\sum_{i=1}^{N} \sigma_{li}\alpha + \sum_{j=1}^{m} \left(\sum_{i=1}^{N} \sigma_{ji}\sigma_{li} \right) \beta_j = \sum_{i=1}^{N} \sigma_{li} A_i^{exp};$$

$$l = 1, 2, \ldots, m$$

The optimisation of coefficients in correlational equations which contain also quadratic terms in σ_{ij} can be performed according to the same principles.

APPENDIX 5

Calculation of the $p(n-\bar{\nu}+1)$ and $\pi(n, \nu)$ probabilities (section 6.4.3)

As mentioned, $p(n - \bar{\nu} + 1)$ is the probability that a n-long randomised array of α-types of monomers has the same monomers in at least $n - \bar{\nu} + 1$ positions as the series corresponding to the combining site of B_0. The relative content, in the whole set of B species, of the monomer found in the position "k" of the B_0 combining site is marked by f_{jk}; $k = 1, 2, \ldots, n$ but $jk = 1, 2, \ldots, \alpha$. The probability to find the same monomers, as in B_0, in at least $n - \bar{\nu} + 1$ fixed positions is a product π over the $n - \bar{\nu} + 1$

f_{jk}-factors. The probability to find the same monomers as in B_0 in less than $n - \bar{\nu} + 1$ fixed positions is one, minus the previous product, $1 - \pi$. The probability to find the same monomers as in B_0, in less than $n - \bar{\nu} + 1$ arbitrary positions out of a total of n is another product: $\pi'(1 - \pi)$, over $C_n^{n-\bar{\nu}+1}$ factors of the type $1 - \pi$. Finally, the probability to find at least one combination (out of a total of $C_n^{n-\bar{\nu}+1}$) of a least $n - \bar{\nu} + 1$ positions (out of a total of n) with the same monomers as in the corresponding positions of the B_0 combining site will be one minus the previous product of products, i.e.

$$p(n - \bar{\nu} + 1) = 1 - \prod_{(jk)}^{(C_n^{n-\bar{\nu}+1})} \left(1 - \prod_{(k)}^{n-\bar{\nu}+1} f_{jk}\right) \cong$$

$$\cong \sum_{(jk)}^{(C_n^{n-\bar{\nu}+1})} \prod_{(k)}^{n-\bar{\nu}+1} f_{jk}$$

The last approximation is justified as, according to inequality (6.14) of section 6.4.3, $p(n - \bar{\nu} + 1)$ must be very small. This is an approximation of the type:

$$1 - \prod_{i=1}^{N}(1 - x_i) \cong 1 - \left(1 - \sum_{i=1}^{N} x_i + \sum_i \sum_j x_i x_j - \ldots\right) \cong$$

$$\cong \sum_{i=1}^{N} x_i \quad \text{for} \quad x_i \ll \frac{1}{N}$$

If all the relative contents are equal, $f_{jk} = 1/\alpha$, one obtains

$$p(n - \bar{\nu} + 1) = C_n^{n-\bar{\nu}+1} \cdot \alpha^{-(n-\bar{\nu}+1)}$$

which introduced into inequality (6.14) gives formula (6.17).

The probability $\pi(n, \nu)$ that a randomized array of length n, formed out of α monomer types should differ in ν positions, as monomer types, from the array corresponding to the combining site of B_0, can be obtained by a similar reasoning. Thus, the probability to have in the positions $1, 2, \ldots, n-$ the same monomer types as in B_0 and in positions $n - \bar{\nu} + 1, \ldots$ $\ldots, n$, different monomers, will be:

$$f_{j1} \cdot f_{j2} \cdot \ldots \cdot f_{jn-\nu} \cdot (1 - f_{jn-\nu+1}) \cdot \ldots \cdot (1 - f_{jn}).$$

Finally one obtains:

$$\pi(n, \nu) = 1 - \prod_{(jk)}^{(C_n^\nu)} \{1 - [(1 - f_{j1}) \cdot \ldots \cdot (1 - f_{j\nu}) \cdot$$

$$\cdot f_{j\nu+1} \cdot \ldots \cdot f_{jn}]\} \cong \prod_{(jk)}^{(C_n^\nu)} [(1 - f_{j1}) \cdot \ldots \cdot (1 - f_{j\nu}) \cdot$$

$$\cdot f_{j\nu+1} \cdot \ldots \cdot f_{jn}]$$

The product and the sum respectively are over C_n combinations of $n - \nu$ positions with the same monomers as in B_0 and positions with different monomers. With equal relative contents, $f_{jk} = 1/\alpha$:

$$\pi(n, \nu) \cong C_n^\nu \left(1 - \frac{1}{\alpha}\right)^\nu \left(\frac{1}{\alpha}\right)^{n-\nu}$$

APPENDIX 6

Deduction of equations (6.18) and (6.21), section 6.4.3

We begin with inequality (6.17), section 7.4.3:

$$NC_n^{n-\bar{\nu}+1} \ll \alpha^{(n-\bar{\nu}+1)}$$

and use Stirling's approximation:

$$\frac{Nn!}{(n - \bar{\nu} + 1)!\,(\bar{\nu} - 1)!} \cong$$

$$\cong \frac{N\sqrt{2\pi n}\; n^n}{\sqrt{2\pi(n - \bar{\nu} + 1)}\,(n - \bar{\nu} + 1)^{(n-\bar{\nu}+1)} \cdot \sqrt{2\pi(\bar{\nu} - 1)}\,(\bar{\nu} - 1)^{(\bar{\nu}-1)}} =$$

$$= N\left(\frac{1}{1 - \dfrac{\bar{\nu} - 1}{n}}\right)^{(n-\bar{\nu}+1)} \cdot \left(\frac{n}{\bar{\nu} - 1}\right)^{(\bar{\nu}-1)} \cdot$$

$$\cdot \sqrt{\frac{n}{2\pi(n - \bar{\nu} + 1)(\bar{\nu} - 1)}}$$

By taking logarithms one obtains

$$\log N - (n - \bar{\nu} + 1) \log \left(1 - \frac{\bar{\nu} - 1}{n}\right) + (\bar{\nu} - 1) \log \frac{n}{\bar{\nu} - 1} -$$

$$- \frac{1}{2} \log \left[2\pi \left(1 - \frac{\bar{\nu} - 1}{n}\right) (\bar{\nu} - 1)\right] < (n - \bar{\nu} + 1) \log \alpha.$$

Rearranging, one obtains:

$$(n - \bar{\nu} + 1) \log \left[\alpha \left(1 - \frac{\bar{\nu} - 1}{n}\right)\right] > \log N + (\bar{\nu} - 1) \log \frac{n}{\bar{\nu} - 1} -$$

$$- \frac{1}{2} \log \left[2\pi(\bar{\nu} - 1)\left(1 - \frac{\bar{\nu} - 1}{n}\right)\right]$$

and formula (6.18) results.

Starting from inequality (6.20), section 6.4.3:

$$\frac{[B_{tot}]}{[B_0]} \sum_{\nu=1}^{n} C_n^{\nu} \left(\frac{1}{\alpha}\right)^{n-\nu} \left[r\left(1 - \frac{1}{\alpha}\right)\right]^{\nu} \ll p_f$$

one observes that the sum over ν represents the nth power of $1/\alpha +$ $+ r(1 - 1/\alpha)$ with the first term, corresponding to $\nu = 0$, missing. Thus one can write:

$$\frac{[B_{tot}]}{[B_0]} \left[\frac{1}{\alpha} + r\left(1 - \frac{1}{\alpha}\right)\right]^{n} \ll p_f$$

Logarithmation gives:

$$\log \frac{[B_{tot}]}{[B_0]} + n \log \frac{1 + r(\alpha - 1)}{\alpha} < \log p_f$$

and formula (6.21) results.